FUNCTIONALLY GRADED MATERIALS IN THE 21ST CENTURY

A Workshop on Trends and Forecasts

FUNCTIONALLY GRADED MATERIALS IN THE 21ST CENTURY

A Workshop on Trends and Forecasts

edited by

Kiyoshi Ichikawa
Mechanical Engineering Laboratory
Agency of Industrial Science and Technology
Ministry of International Trade and Industry
Japan

KLUWER ACADEMIC PUBLISHERS
Boston / Dordrecht / London

Distributors for North, Central and South America:
Kluwer Academic Publishers
101 Philip Drive
Assinippi Park
Norwell, Massachusetts 02061 USA
Telephone (781) 871-6600
Fax (781) 871-6528
E-Mail <kluwer@wkap.com>

Distributors for all other countries:
Kluwer Academic Publishers Group
Distribution Centre
Post Office Box 322
3300 AH Dordrecht, THE NETHERLANDS
Telephone 31 78 6392 392
Fax 31 78 6546 474
E-Mail <orderdept@wkap.nl>
Electronic Services <http://www.wkap.nl>

Library of Congress Cataloging-in-Publication Data

Functionally graded materials in the 21st century : a workshop on trends and forecasts/
edited by Kiyoshi Ichikawa
p. cm.
ISBN 0-7923-7236-0
1. Functionally gradient materials--Congresses. I. Ichikawa, Kiyoshi,1943-

TA418.9.F85F89 2000
620.1'18--dc21

00-048693

Printed on acid-free paper.

Printed in the United States of America

Contents

Organizing Committee

Dr. N. Ooyama

Director-General of Mechanical Engineering Laboratory, Agency of Industrial Science and Technology, Ministry of International Trade and Industry

Dr. K. Ichikawa

Chief of Materials Design Division, Mechanical Engineering Laboratory, Agency of Industrial Science and Technology, Ministry of International Trade and Industry

Dr. M. Koizumi

President of FGM Forum, Professor Emeritus, Supervisor of REC, Ryukoku University

Dr. T. Hirai

Professor & University Councilor of Tohoku University, Director of Laboratory for Developmental Research of Advanced Materials of Institute for Materials Research

Dr. Y. Miyamoto

Professor of Materials Science at The Research Center for Cyclic Loop System for Processing and Maintenance, Joining and Welding Research Institute, Osaka University

Dr. M. Niino

Chief of Rocket High Altitude Performance Section, Kakuda Research Center, National Aerospace Laboratory, Science Technology Agency

M. Tokita

Managing Director of SCM Division, Sumitomo Coal Mining Group, Izumi Technology Company Ltd.

Dr. A. Yabe

Chief of Research Planing Office, Mechanical Engineering Laboratory, Agency of Industrial Science and Technology, Ministry of International Trade and Industry

Contributors

S. Alam, S. Sasaki and H. Shimura (*3.2.3*)

Mechanical Engineering Laboratory, *Tribological and Metallurgical Evaluation of Molybdenum with Copper Composite Material Produced by Laser Assisted Low Pressure Plasma Spraying Technique*

S. Amada (*4.7.3*)

Gunma University, *Hierarchical and Functionally Graded Structure of Plants*

H. Awaji, H.Takenaka, S. Honda and T. Nishikawa (*2.2*)

Nagoya Institute of Technology, *Analysis of Temperature and Stress Distributions in a FGM Plate under Thermal Shock*

T. E. Fischer, Y. F. Qiano and Z. Zhu (*4.6.1*)

Stevens Institute of Technology, *Tribological Approach to Functionally Graded Materials*

Y. Hibi[1], Y. Enomoto[2] and A.Tanaka[1] (*4.6.3*)

[1]Mechanical Engineering Laboratory, [2]Nagoya National Industrial Research Institute, *Evaluation of Wear Resistance of Si_3N_4-TiN-Ti Composites*

T. Hirano, J. Teraki and Y. Nishino (*4.5.4*)

Daikin Industries Ltd., *Computational Design for Functionally Graded Thermoelectric Materials*

K. Ichikawa[1] and S. Miyamoto[2] (*4.4.4*)

[1]Mechanical Engineering Laboratory, [2]Izumi Technology Company, Ltd., *Tensile Properties of ZrO_2(3Y)/Stainless Steel FGMs Fabricated Spark Plasma Sintering Method*

N. Ichinose (*4.5.1*)

Waseda University, *Preparation and Properties of Functionally Graded Electronic Materials*

A. Ikegaya[1], K.Tsuda[1], T. Miyagawa[2] and Y. Suehiro[3] (*5.1*)

[1] Sumitomo Electric Industries, Ltd., [2] JRCM, [3] Japan National Oil Corporation, *Application of Functionally Graded Materials for Tool Materials*

S. Kano (*4.2.2*)
Mechanical Engineering Laboratory, *Residual Stress of ZrO_2/SUS FGM by XRD*

M. Kawahara (*4.4.3*)
Izumi Technology Company, Ltd., *Properties of Large-Size ZrO_2(3Y)/ Stainless FGMs Fabricated by Spark Plasma Sintering Method*

A. Kawasaki (*4.1*)
Tohoku University, *Evaluation Methods of Functionally Graded Materials*

M. Kitayama[1], T. Narushima[2] and **A.M. Glaeser**[3] (*2.4*)
[1] FCRA, Nagoya, [2] Tohoku University, [3] University of California, *Tools for Model Studies of the High-Temperature Properties of Surfaces and Interfaces*

Y. Miyamoto[1], J. Oh[1], S. Kiriharra[1] and K. Matsuura[2] (*1.2*)
[1] Osaka University, [2] Hokkaido University, *Cybermaterials: Concept, Processes and Applications*

K. Mizuhara[1], K. Tanaka[2] and N. Morita[2] (*4.2.3*)
Mechanical Engineering Laboratory, [2]Chiba University, *Evaluation of Residual Stresses in FGM by Means of Optical Methods*

T. Murakami (*4.3*)
[1]Mechanical Engineering Laboratory, *Oxidation Behavior of Spark Plasma Sintered Compacts of the Nb-Si-X (X:B, Cr, Al) Systems*

Z. A. Munir (*3.1.1*)
University of California, *One-Step Synthesis and Densification of Monolithic and FGM Materials by Electric Field Activation*

T. Nakamura and S. Sampath (*3.2.2*)
State University of New York at Stony Brook, *Manufacturing of Thermally Sprayed FGM and Determination of Its Property by Inverse Analysis*

Y. Nakayama (*3.1.4*)
Izumi Technology Company, Ltd., *Development of Automatic FGM Manufacturing System*

M. Niino and S. Moriya (*5.3*)
National Aerospace Laboratory, *Launch into Space with FGM*

Y. Okazaki (*4.7.2*)
Mechanical Engineering Laboratory, *Surface Modification of New Ti Alloy for Medical Application*

M. Omori and T. Hirai (*3.1.2*)
Tohoku University, *Fabrication of FGM Using the Function of SPS*

M. Pindera[1], J. Aboudi[2] and S. M. Arnold[3] (*3.2.1*)
[1]University of Virginia, [2]Tel-Aviv University, [3]NASA-Glenn Research Center, *Modeling Issues for Plasma-Sprayed TBCs with Graded Bond Coat Architectures*

E. Reimanis[1], J. Chapa[1], A. N. Winter[2], W. Windes[2] and E. Steffler[2] (*4.4.1*)
[1]Colorado School of Mines, [2]Idaho National Engineering and Environmental Laboratory, *Fracture and Deformation in Ductile/Brittle Joints with Graded Structures*

K. Shinagawa (*2.1*)
Anan College of Technology, *Computer Aided Design of Graded Powder Compacts to Control Sintering Defects*

D. T. Smith (*4.4.2*)
National Institute of Standards and Technology, *Use of Instrumented Indentation to Determine Mechanical Properties of Thin Films, Coatings and Functionally Graded Materials*

S. Takatsu and K. Umeda (*4.6.2*)
Mechanical Engineering Laboratory, *Tribological Properties and Damage of ZrO_2 Layer in ZrO_2-Stainless Steel FGM with a Pin on Block Friction*

J. Takatsubo[1], M. Imade[1], S. Yamamoto[1], Y. Enomoto[2] and M. Tokita[3] (*4.2.1*)
[1]Chugoku National Industrial Research Institute, [2]Nagoya National Industrial Research Institute, [3]Izumi Technology Company, Ltd., *Ultrasonic Imaging of Metal/Ceramics FGM*

M. Tokita (*3.1.3, 5.2*)
Izumi Technology Company, Ltd., *Advanced SPS Systems for FGM Technology, Industrial Applications for Functionally Graded Materials Fabricated by Spark Plasma Sintering (SPS) Systems*

T. Tsujikami, M. Kawagai, K. Inoue and M. Ohyanagi (*2.3*)
Ryukoku University, *Evaluation of Residual Stress for Particle Gradually Dispersed Composites by Finite Element Method*

S. Uemura[1], M. Niino[2] and M. Koizumi[3] (*1.3*)
[1]Kyusyu University, [2]National Aerospace Laboratory, [3]Ryukoku University, *The Activities of FGM Forum*

R. Watanabe[1], T. Nishida[2] and T. Hirai[1] (*1.1*)
[1]Tohoku University, [2]Kyoto Institute of Technology, *Present Status of Research on Design and Processing of Functionally Graded Materials*

S. Watanabe[1], R. Yamashita[1], S. Katoh[1], N. Hayashi[1], Y. Uchida[1], S. Higa[1], D. Dykes[2] and G. Touchard[3] (*4.5.2*)
[1]Aichi Institute of Technology, [2]Yokkaichi University, [3]Universite de Poitiers, *The Crystalline Structure and Electrical Properties of TiO_2-based Functionally Graded Materials*

Y. Watanabe (*3.2.4*)
Shinsyu University, *Magnetically Graded Materials by Martensitic and Reverse Martensitic Transformation Techniques*

F. Watari[1], A. Yokoyama[1], H. Matsuno[1], R. Miyao[1], M. Uo[1], T. Kawasaki[1], M. Omori[2] and T. Hirai[2] (*4.7.1*)
[1]Hokkaido University, [2]Tohoku University, *Fabrication of Functionally Graded Implant by Spark Plasma Sintering and Its Biocompatibility*
X. Wang[1], L. Chen[1], T. Hirai[1] and Y. Someno[2] (*4.5.3*)
[1]Tohoku University, [2]ALPS Electric Co., Ltd., *Design of High Performance Reflection Filter of SiO_2-ZrO_2 Multilayer Films with Graded Refractive Index Profiles*

Preface

I am honored to chair this International Workshop on Functionally Graded Materials in the 21st Century: A Workshop on Trends and Forecasts, and would like to first express my sincere gratitude to everyone participating. The Mechanical Engineering Laboratory and the Japan International Science and Technology Exchange Center (JISTEC) have co-organized this workshop with the sponsorship of the Science and Technology Agency of Japan and the cooperation of the Association of Mechanical Technology. This workshop is an international conference to focus on functionally graded materials and the aim is to provide an overview of the present global technical trends and the future development of functionally graded materials over the next 10 years. I am very happy to see many researchers meeting together here — including seven researchers invited from abroad. During the three-day oral sessions, 36 research reports will be presented, and I'm sure I'm not the only one who is very anxious to hear and participate in the upcoming interesting discussions.

At present, the Mechanical Engineering Laboratory is conducting fundamental and ground-breaking research in such major areas as materials science and technology, bioengineering, information & system science, advanced machine technology, energy technology, manufacturing technology and robotics.

In particular, we consider research on materials science and technology to have the highest priority for the 21st century, and since 1996 have participated in the US-Japan joint research project, *Precompetitive Processing and Characterization of Functionally Graded Materials.*

The Mechanical Engineering Laboratory of Japan and the US National Institute of Standards have been carrying out this 4-year joint research project to produce and evaluate new functionally graded materials using spark plasma sintering. In Japan, Sumitomo Coal Mining Group, Izumi Technology, Co. Ltd. and Yanmar Diesel Engine Co. have joined the project. This joint research focuses on production of large-scale functionally graded material composites of a metal with high mechanical strength and a ceramic with high heat-resistance and high-wear resistance.

At present, international conferences related to functionally graded materials are being held all over the world, and many notable research achievements in processing and evaluation have been reported.

In light of the current development efforts on functionally graded materials at research institutes throughout the world, the objective of this workshop is to provide researchers an opportunity to present their respective accomplishments and exchange valuable information with their colleagues in the field.

I hope you all will benefit from the wide array of presentations scheduled for the workshop, including those on FGMs for modeling and simulation, coating, bulk processing, properties and characterization. I also hope you will discuss the value and necessity of continuing to hold workshops such as this one, and I hope this workshop will be the start of many more to come. I would like to conclude my address by once again thanking everyone here for your participation.

Dr. Naotake Ooyama
Chairman, Functionally Graded Materials in the 21st Century: A Workshop on Trends and Forecasts
Director-General, Mechanical Engineering Laboratory
Agency of Industrial Science and Technology, MITI

Acknowledgements

All the authors would like to gratefully acknowledge the sponsorship of the Science and Technology Agency, Japan, for the international workshop, *Functionally Graded Materials in the 21st Century: A Workshop on Trends and Forecasts,* which was held on March 26 to 28, 2000, at Epochal Tsukuba International Congress Center, Tsukuba, Japan.

The studies in Sections 1.1 and 1.3 were supported by a Grant-in-Aid for Scientific Research on a Priority Area from the Ministry of Education, Science, Sports and Culture of Japan. The study in Section 2.1 was supported by the Amada Foundation for Metal Work Technology. The sintering experiment was conducted by Dr. Hirashima at the Tokushima Prefectural Industrial Technology Center. The study in Section 3.1.1 was supported by a grant from the Army Research Office. The authors of Section 3.1.2 acknowledge the support of the Ministry of Education, Science and Culture under a Grant-in Aid for Scientific Research on Priority Areas (B) (No. 11221203, Harmonic Material Design of Multi-Functional Composites), and also appreciate being permitted to use the spark plasma sintering system managed by the Laboratory for Developmental Research of Advanced Materials, Institute for Materials Research, Tohoku University. Additionally, the authors wish to thank Mr. Y. Murakami for SEM and EPMA. The Cu, thermosetting polyimide, WC, Mo and Co powders were supplied by Nippon Atomized Powders Co. Ltd., Kaneka Co. Ltd., Japan New Metals Co. Ltd., Tokyo Tungsten Co. Ltd., and Kokkola Chemicals, respectively.

The studies in Sections 3.1.1, 3.1.4 and 4.2.2 were conducted as a part of the Development of Functionally Graded Materials Project sponsored by NEDO. The authors of Section 3.2.2 acknowledge the support of NSF under grant CMS-9800301. The authors of Section 3.2.3 wish to thank Dr. S. Kano and Ms. K. Saito for their valuable technical support, and the New Energy Development Organization (NEDO) for financial assistance to one of the authors. The author of Section 3.2.4 would like to thank Prof. Y. Fukui, Prof. Y. Nakamura and Prof. K. Nakanishi at Kagoshima University, and Prof. J. W. Morris, Jr, Dr. J. W. Chan and Dr. S. H. Kang at Lawrence Berkeley National Laboratory for their assistance in this study. Portions of this study were supported by Grant-in-Aid for COE Research (10CE2003) by the Ministry of Education, Science, Sports and Culture of Japan. The author of Section 4.2.3 would like to express his appreciation to Dr. S. Takatsu of MEL for his assistance in preparing the FGM samples. The study in Section 4.4.3 was performed as a part of the Development of Functionally Graded Materials Project sponsored by the New Energy and Industrial Technology Development Organization (NEDO). The author wishes to acknowledge the support of the Mechanical Engineering Laboratory (MEL), the Agency of Industrial Science and Technology, the Ministry of International Trade and Industry (MITI), NEDO and the Technical Research Center of Yanmar Diesel Engine Company.

Portions of the study in Section 4.5.3 were supported by (1) Grant-in-Aid for Encouragement of Young Scientists (A) (No. 11750578) and a Grant-in-Aid for Scientific Research (A) (No. 11355027) from the Japan Society for the Promotion of Science. (2) Grant-in-Aid for Scientific Research on Priority Areas (No. 08243102) from the Ministry of Education, Science, Sports and Culture, Japan. (3) Grant-in Aid for the Promotion of Research of Advanced Materials (No. H1001p) from the Laboratory for Developmental Research of Advanced Materials, Institute for Materials Research, Tohoku University. The authors of Section 4.5.4 would like to thank Prof. K. Masui for providing the experimental data on BiSb electrodeposition prior to publication. A portion of this study was supported by a grant for Research and Development of Advanced Computational Science and Technology from Japan Science and Technology (ACT-JST project). The authors of Section 4.6.1 gratefully acknowledge the financial support of the Office of Naval Research under Grant Number N00014-97-1-0844.0. The authors of Section 4.6.3 wish to thank Mr. H. Ikebe for his help with the experiments.

Finally, all the authors would like to thank Dr. Y. Enomoto, the former Deputy-Director-General of Mechanical Engineering Laboratory for the support to the international workshop and Ms. M. Yamauchi for the assistance to the editor.

Chapter 1

INTRODUCTION

1. STATE OF THE ART

1.1 Introduction

Direct bonding of metals and ceramics can generate thermal stresses in practical high-temperature applications which are due mainly to CTE mismatches. The thermal stress causes crack formation, debonding at hetero-interfaces and often results in delamination of the ceramic over-layer. A graded interlayer concept for metal/ceramic bonding was first proposed by Kawasaki and Watanabe [1], in which the thermal stress was effectively eliminated by a compositionally graded interlayer. Such graded materials possess simultaneously a super-heat-resistant property and sufficient toughness to arrest crack propagation. This concept was given the name of functionally gradient material (FGM) [2] and was originally applied to create the fuselage exterior and engine materials for space planes which would take off like airplanes, cruise at Mach 5 to 25 in the atmosphere subject to severe frictional heating from the airflow. Optimal response of material properties to conditions in an actual environment is the main requirement in the design of FGMs. To fulfill this requirement the composition and microstructure are varied throughout the structure, and this yields a property gradient within the combined materials. The FGM concept was also applied to improve the figure of merit of thermoelectric materials. In this case the carrier concentration is graded by finely control the doping element concentration, thus causing a wide temperature range with a high figure of merit [3]. The most remarkable difference between FGMs and conventional homogeneous materials lies in their needs-oriented materials design and non-homogeneous microstructural control. Such a concept is applicable to functional materials in various fields, wherein several functions are incorporated in a single material or new functions are expected to appear due to the compositionally graded structure [4]. Another significant advance in FGM research was brought about by the recently finished Physics and Chemistry of Functionally Graded Materials project (FY 1996-1998) supported by Japan's Ministry of Education. A great variety of processing approaches were proposed to manifest various material functions, and they included the sol-gel method, infiltration, polymerization, substitution reaction, eutectic phase reaction, electropholetic method, hydrothermal-electrochemical method, and molten salt reaction. Table 1.1 is a summary of the research topics in the project, which cover physical, chemical, agricultural and medical functions [5]. Not only composition as in conventional usage but also various microstructural features and electronic structures were graded, and the modifications of the material properties were investigated in detail.

Table 1.1. Research categories and number of topics in the Monbusho Project (1996-1998), Physics and Chemistry of Functionally Graded Materials.

Theme		Year	1996	1997	1998
Physical function	Single function	Theory	3	3	2
		Mechanical	7	6	6
		Electrical	9	7	4
		Magnetic	5	6	3
		Optical	4	7	4
	Multi-function	Thermal - Electrical	5	5	5
		Electrical - Magnetic	1	1	1
		Magnetic - Optical	1	1	1
		Electrical - Optical	4	4	5
		Electrical - Strain	2	2	1
Chemical function		Chemical reaction	8	8	6
		Catalyst and sensor	4	4	1
		Separation film	3	1	1
		Corrosion resistance	1	4	2
		Battery	3	1	2
Agricultural function			3	3	2
Medical function			5	9	6

In this section several typical examples of successful applications of the FGM concept are briefly reviewed, and the results of research on the design and processing of FGMs from the recently completed Monbusho Project [5] are given, in which some selected topics with well-defined design concepts are described in further detail.

1.2 Practical Application Examples

1.2.1 Thermal Barrier/Anti-Oxidation Coatings

Refractory ceramics are coated on a metallic substrate by inserting a graded interlayer to assure good bonding between the two materials and to reduce the thermal stresses that would otherwise be generated due to large CTE mismatches at the interfaces. Zirconia coating on a nickel-chromium alloy substrate by plasma spraying [6] and SiC coating on C/C composites by CVD [7, 8] are typical examples of practically applied FGM coatings. The composition profiles of the graded interlayer are carefully controlled to optimum compositional profiles that will give the least stress generation. In these cases the requirements are fulfilled for the heat-resistant and oxidation-resistant properties, as well as for adequate mechanical toughness. The FGM coating will be used for gas turbine blades in power generators to improve combustion efficiency, and will surely be one of the most important technologies for saving energy in near future, thus helping to lower emissions into the environment. Long thermal fatigue life is presently an emerging subject to be investigated in these FGM coatings. The basic mechanism of crack formation and propagation leading to the delamination of the overlayer has been clarified by Kawasaki and Watanabe [9] through burner heating test results and by the fracture mechanical approach. The thermal fatigue properties of thermal barrier coatings have been extensively investigated worldwide [10-12].

1.2.2 Cutting Tools: Cemented Carbides

Sumitomo Electric Industries Ltd. developed FGM cemented carbide with improved wear resistance and long tool life [13]. In this example TiCN-WC-Co-Ni alloy is sintered in a controlled nitrogen atmosphere and at a controlled cooling rate to create a TiC-TiCN-WC graded microstructure with Co as a binder phase. The uppermost TiC Phase is a harder phase and has a good anti-wear property and the inner WC-rich phase has good toughness. The surface region itself contains no Co binder but the binder content increases from just beneath the TiC overlayer towards the inner region. This causes a CTE gradation in the surface region, generating compressive stress in the TiC overlayer that enhances fracture toughness and prolongs tool life.

Due to the graded structure the bonding of the TiC and the core material is better than in conventional TiC coatings on WC-Co alloy. Another type of FGM cemented carbide was developed by Honda Engineering Co. Ltd. [14]. A mixture of 98%WC-2%TaC-1%NbC-8%Co was pre-sintered to form a porous compact, impregnated with Ni solution (equivalent Ni content of 2%), and then sintered in nitrogen or in vacuum. The surface region has a coarse grain size and less cobalt content than in the core region with the graded interlayer. The hardness is highest in the surface layer, and the toughness increases towards the core region. These microstructural and property gradations endowed this material with good cutting performance and long life.

1.2.3 Cutting Tools: Diamond/SiC

Japan Ceratec Co. Ltd. and Tohoku University developed a diamond-chip/SiC-shank tool for precision machining that possesses high stiffness and long tool life [15]. This FGM tool has a sound bonding interface with no cracks near the solder-bonded interface as are usually found in directly bonded diamond tools. The successful bonding of the diamond chip and the SiC shank was achieved by inserting a graded interlayer and using conventional reaction sintering of the porous SiC body. The composition profile was strictly calculated by FEM and using measured data on the compositional dependence of the thermal and mechanical properties. According to the FEM analysis the residual stresses that are generated during cool-down from the processing temperature completely disappear in the bonded and graded regions, and only a low-level thermal stress, which is small enough compared to the strength of the reaction bonded SiC, remains in the shank. The machining accuracy was improved about 6-fold without wavy vibration and the tool life was greatly prolonged compared to conventional diamond tools.

1.2.4 Thermoelectric Materials

Nishida [3] proposed an improvement of the figure of merit by grading carrier concentration or dopant content. The figure-of-merit vs. temperature curve shifts towards the high-temperature side depending on the carrier concentration. If the carrier concentration is changed gradually within a single material, the peaks for each concentration value are bound to form a envelope with a large figure-of-merit and high power output. A similar principle can be applied to a combination of different of thermoelectric materials which exhibit different optimum temperature ranges. Hirano [16] verified this prediction with a layered material of BiTe/PbTe/SiGe bonded by graded layers.

1.2.5 Optical Film: Bandpass Filter

The Hirai group of IMR at Tohoku University [17, 18] proposed a refractive-index-graded optical film, in which the refractive index is graded to yield perfect light transmittance or reflectance depending on the wavelength. They fabricated the graded SiO_2-TiO film by helicon sputtering according to a theoretical index profile, and verified that the filter material exhibited high reflectance or transmittance (>99%) in the very narrow band. This FGM optical film will be used as a bandpass filter in place of a conventional rugate filter.

1.2.6 Piezoelectric Actuator

The Watanabe group of the School of Engineering at Tohoku University [19, 20] proposed a porosity-graded piezoelectric material that exhibits a large bending deflection as in conventional unimorph actuators. Micromechanical analysis has shown that an optimum porosity distribution profile exists which gives maximum deflection. According to the theoretical design they fabricated a rectangular-plate-shaped material, and the measured deflection agreed well with the forecast values.

1.3 Brief Review of the Monbusho Project

The most recent national project on FGM in Japan was carried out by university researchers with the support of the Ministry of Education and was entitled Physics and Chemistry of Functionally Graded Materials (FY 1996-1998). The project leader was Professor Toshio Hirai of IMR at Tohoku University. Watanabe of Tohoku University joined this project as a contributing member and helped as a coordinator of the processing group.

Tables 1.2(a) and (b) summarize the research topics, in which new processing methods are employed to realize the target functions. Process routes, gradients, materials, functions and applications, investigators and additional remarks were tabulated. Table 1.2(a) lists seven selected topics in FGM processing which were conducted with overriding priority. In these investigations the design and processing methodology for the compositional and microstructural grading according to FGM design were studied. Table 1.2(b) lists secondary work by other groups that involved physical, chemical, medical and biological functions, but employed original processing methods to prepare their samples. While detailed explanations of these topics will be omitted in Table 1.2(b) and only some of the topics listed in Table 1.2(a) will be reviewed, it is hoped that readers will see that there are as many process routes as there are different functions.

1.3.1 Spark Plasma Sintering (SPS)

Omori and Hirai of Tohoku University employed SPS to consolidate polyimide/Cu graded materials. SPS is a pressure-assisted and rapid process, and is suitable for the consolidation of hard-to-sinter powdered materials.

Table 1.2(a). Processing routes and applications of functionally graded materials investigated in the Monbusho Project (1996-1998). (From the leading group of design and processing)

Process	Gradient/ Materials/ (Thickness)	Functions & Applications	Remarks (Author)
Spark plasma sintering	Composition/ Polyimide/Cu Hydroxyapatite/Ti WC/WC-Co/ (Seveal mm)	IC substrate Artificial bone Cutting tools	(Omori & Hirai: TohokuUniv.)
Particle assembling	Composition/ Steel balls (model material) BiSb/ (Several mm)	Thermoelectric device	(Watanabe: Tohoku Univ.)
Combustion synthesis	Composition/ TiB_2-Si/diamond particles/ (Several mm)	Heat sink Cutting tool	(Ohyanagi: Ryukoku Univ.)
Sol-gel Process with repeated dip-coating	Composition (solid solution)/ $Ti_{1-x}V_xO_2$ (x=0-0.1)/ (sub microns)	Wide range of visible light photoresponces Solar cell electrode	(Yoko: Kyoto Univ.)
Plasma spraying	Doping element Carrier-concentration/ Si-Ge-P(1-3 at%)/ (several mm)	Thermoelectric performance	(Fukumoto: Toyohashi Univ. of Technology)
Sedimentation of atoms under a million g	Atomic composition/ Bi-Sb, Bi-Pb Bi-Sn/ (Several mm)	(Mezo-scale control)	(Mashimo: Kumamoto Univ.)
Precursor process	Composition/ Mo - Si - C - N/ (Several mm)	Oxidation resistance coating for C/C composite	(Sasaki: Muroran Inst. Technology)

Table 1.2(b). Processing routes and applications of functionally graded materials investigated in the Monbusho Project (1996-1998). (Selected from all topics in the project)

Process	Gradient/ Materials (Thickness)	Functions & Applications	Remarks (Author)
Infiltration into porous preform	Composition/ $BaTiO_3$-Si / (1 mm)	Dielectric material Low signal loss at high frequency	Si:0.5-4.5 mass% (T.Nishida:Kyoto Inst. Tech.)
Spray deposition into polymer foam	Composition and porosity/ Cordierite/Mullite/ (2 mm)	Thermal shock resistance Filter for coal - fueled generator	Two-nozzle spraying (K.Kodaira: Hokkaido University)
Laser CVD (Photochemical reaction)	Composition/ SiC/ C / (1 mm)	Oxidation resistance coating	Low temperature CVD (O.Ito:Tohoku University)
Cast and diffusion (Miscible polymer blend)	Composition/ PMMA / SAN polymers film/ (Several mm)	Gas permeation membrane	Cast into two layers and diffusion-treated (Y.Tsujita:Nagoya Inst. Tech.)
Electropholetic deposition (Colloidal process)	Composition/ aFe_2O_3 / YSZ Fe_2O_3 / $ZnFe_2O_4$/	Ceramic coating	PH (ζ potential) control (F.Utsuno:Univ. of Tokyo)
Self-structuring and dissolution process	Composition/ PVA / PVP PMMA / PEG PP / PEG PMMA / PVAc/ (Several mm)	Super large surface area	Self-organizing phase separation in polymer blend (Yamagata Univ.)
Vulcanization and cure	Cross-linking density/ Butyl rubber/ (2 mm)	Improved damping capacity	Valcanization-layering-pressing-curing (Y.Ikeda:Kyoto Univ.)
Pack-cementation	Phase and Composition/ Ti-43Al-5Cr /L1 intermetallics/ (70 mm)	Oxidation resistant	Pack powder:Ti-43Al-5Cr $+Al_2O_3$ (H.Mabuchi:Osaka Prefecture Univ.)

PMMA:Polymethyl metha-crylate, SAN:Stylene-co-acrylonitrile, PVA:Polyvinyl alcohol, PVP:Polyvinyl pyrolidone, PEG:Polyethylene glycohol, PP:Polypropylene, PVAc:Polyvinyl acetate

Table 1.2(b). (Continued)

Process	Gradient/ Materials (Thickness)	Functions & Applications	Remarks (Author)
Filtration-Sintering	Porosity/ Hydroxyapatite/ (Several cm)	Biocompatible materials	HAp+Polymer suspension- (M.Iwata:Suzuka National College of Tech.)
Centrifugal forming of semi-solid composite	Composition/ Al / Al_3Ni Al / Al_3Fe/ (Several mm)	Wear resistant materials	0 - 50 Vol% intermetallics in metal matrix (Y.Fukui:Kagoshima University)
Eutectic reaction	Phase & composition/ Ti/Ti_3Sn, Ti/Ti_5Si_3 Ni/TiC, Ni/TaC (3 mm)	Heat resistant materials	(T.Tsujimoto:Ibaraki University)
Substitutional reaction	Al/Al_2O_3	Anti-wear machine parts	Substitutional reaction between SiO_2 and Al melt (N.Yoshikawa: Tohoku Univ.)
Hydrothermal-electrochmeical synthesis	$BaTiO_3$/$SrTiO_3$/Ti $SrTiO_3$/$BaTiO_3$/Ti	DRAM capacitor	Synthesis temp. <200C (M.Yoshimura: Tokyo Inst. Tech.)

The sintering is enhanced by electric spark discharge. As shown in Figure 1.1 a slanted graphite die was used to give the sample a temperature gradient, thereby giving different temperatures to the polyimide and copper powder, both of which have different sintering temperatures. The assumed application of this FGM is an IC substrate with high thermal conductivity and electrical insulation. This method was successfully applied to consolidate hydroxiapatite/Ti and WC/WC-Co FGMs [21].

1.3.2 Particle Assembly

Watanabe *et al.* [22] devised a particle assembly system using a suction tip to pick up a particle and convey it to an allotted cite at a two-dimensional lattice point (Figure 1.2). Two-dimensional particle arrays of two different particle species are constructed for each of the mixing ratios according to a random particle configuration. These arrays with gradually varied mixing ratio are stacked layer by layer to form a three-dimensional, composition-graded particle assembly. This system, together with specially manufactured mono-sized spherical particles [23], allows the fabrication of particle assemblies of any random and regular particle configurations. Watanabe intended to use this assembly to verify the percolation control theory to improve the figure of merit of graded thermoelectric composites [24]. This method, on the other hand, can be applied to fabricate micro-machines and micro-devices from functional particles, in which the particles are arranged and bonded to each other to form minute structures.

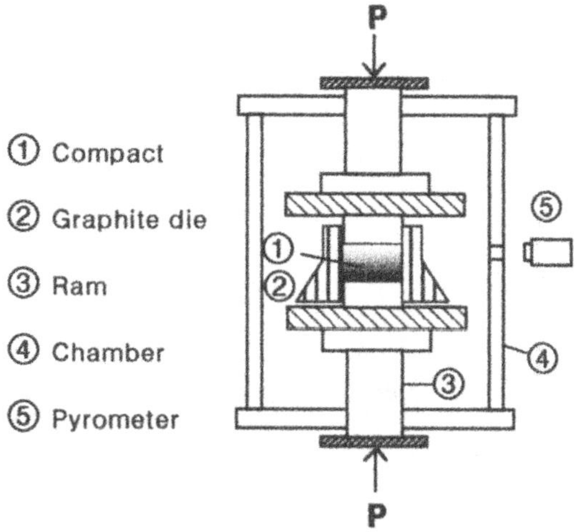

Figure 1.1. Spark plasma sintering of functionally graded materials with a slanted graphite die.

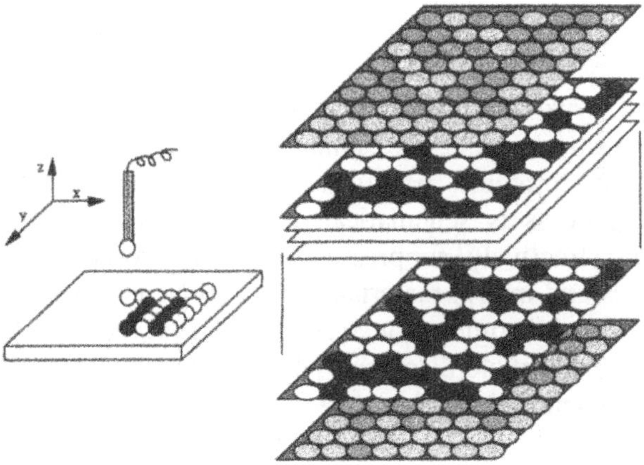

Figure 1.2. Schematic illustration of a particle assembly method to form a three-dimensional graded structure.

1.3.3 Plasma Spraying

Fukumoto of the Toyohashi University of Technology carried out a detailed investigation of the microstructure and property control of plasma sprayed films. Plasma spraying is a fast process and provides film formation over a wide area, but usually has a disadvantage with its rough microstructure. Fukumoto used mechanically alloyed powders of thermoelectric materials, and succeeded in finely controlling the dopant content in SiGe-P(1-3%) to form a carrier-concentration-graded thermoelectric material with a homogeneous microstructure [25]. His spraying configuration is schematically shown in Figure 1.3.

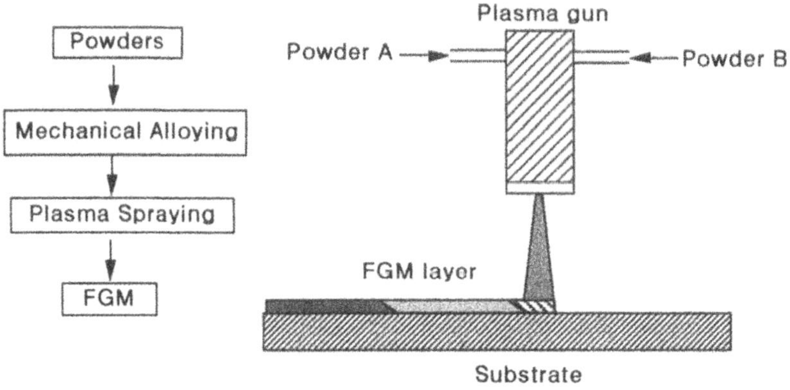

Figure 1.3. Conceptual illustration of a plasma spraying method to form a graded thermoelectric material. (The composition is graded in the plane direction.)

1.3.4 Combustion Synthesis

Ohyanagi of Ryukoku University [26] constructed a timely pressurizing system for combustion synthesis of graded composites. The timely pressure addition results in complete densification which would otherwise not be attainable. The process is schematically shown in Figure 1.4. The short processing time effectively suppresses interfacial reactions between the diamond particles and matrix material. After precise thermal stress analysis and fine control of the combustion reaction, he succeeded in fabricating a heatsink of diamond-particle-dispersed TiB_2-Si cermet and also a cutting tool of WC-Co/diamond particles.

1.3.5 Sol-Gel Process

Yoko of Kyoto University [27] employed repeated dip-coating and

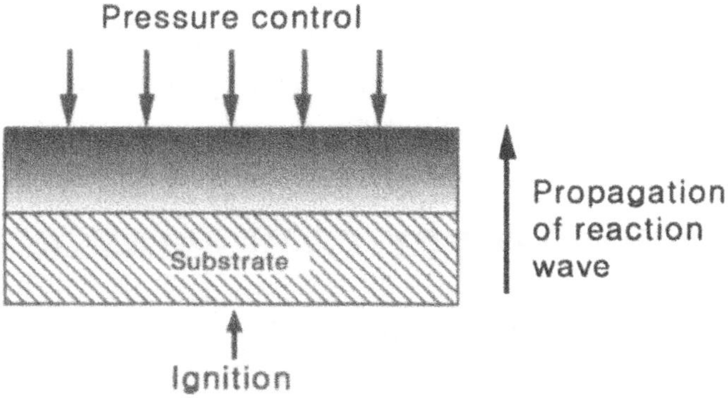

Figure 1.4. Schematic illustration of combustion synthesis of a graded composite. (Pressure is applied just after the combustion wave passed through the sample.)

layered precursor formation of $Ti_{1-x}VxO_2$ (x=0-0.1) on a nesa-silica glass. The layered material was subsequently heat-treated to form a continuously composition-graded solid solution and thus having a graded band-gap semiconductor (Figure 1.5). This film exhibits a wide range of visible light photo-response with photochemical stability and has practical application to the corrosion-resistant film electrode for solar cells.

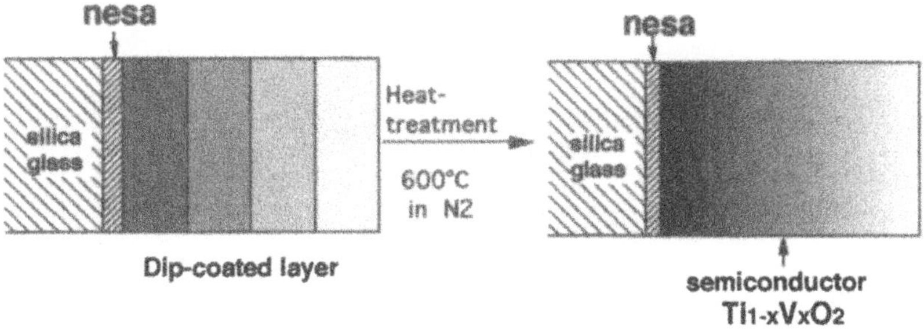

Figure 1.5. Sol-gel process with repeated dip-coating to form a precursor and subsequent heat-treatment to form a continuously graded solid solution of $Ti_{1-x}V_xO_2$.

1.4 Concluding Remarks

A great variety of processing routes is now available for the fabrication of FGMs. The Monbusho Project has revealed that there are many potential application fields for the FGM concept. A careful examination of successful FGM applications reveals that they all have the following features: (1) theoretical design for a target function, (2) optimum size and shape, (3) appropriate processing route, (4) material-property data pertaining to the employed processing route for a realistic design, (5) evaluation of the function under well-defined methodology and its feedback to secondary design. These features would also be a guideline for more effective processing of new FGM applications.

2. NEW CONCEPT: CYBERMATERIALS

2.1 Concept of Cybermaterials

With the rapid development of the Internet, or cybernets, the face of design engineering has changed dramatically. CAD, CAM and other digital data can be viewed and manipulated from remote locations into a three-dimensional model that can then be manufactured almost anywhere. When this approach is extended to rapid prototyping for materials processing, the next generation of materials—cybermaterials—is created. The concept of cybermaterials engineering is illustrated in Figure 1.6.

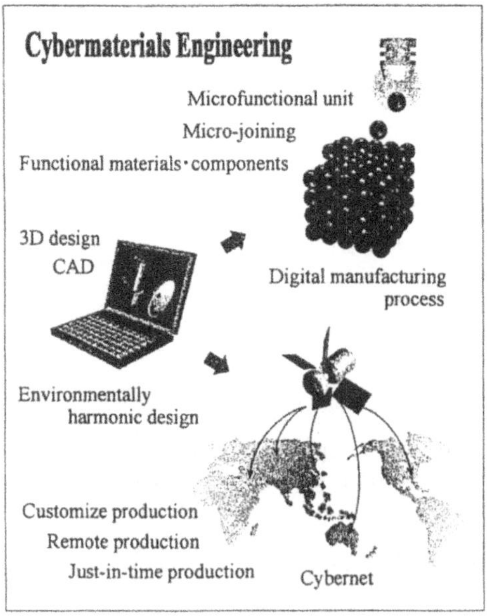

Figure 1.6. Image of cybermaterials engineering.

A cybermaterial can be composed of micro-functional units, which are arranged three-dimensionally by digital-based manufacturing. A micro-functional unit is similar to the concept of functionally graded materials [28, 29] and can range in size from atomic to mm scales. Small liquid droplets, solid particles, or vapor beams can be used to form micro-functional units. Cybermaterials could be processed as follows. First, one can invent almost any type of new material, then design the 3D-microstructure of the micro-functional units using a CAD system. The 3D-design of the material structure is converted to a numerical code file for processing into 3D-microstructures. For instance, micro-functional units can be injected onto an x-y-z stage from a nozzle and built into 3D-microstructures, resulting in freeform fabrication of components with optimized structures, compositions, and functions.

This fabrication technique is known as rapid prototyping[28, 30]. There are many rapid prototyping methods. However, most of these methods are traditionally used for fabricating molds or models. Cybermaterials have active functions in addition to their free-forming capability. Using these methods, one can fabricate macro-functional components or create artificial crystals with periodic structures.

2.2 Expected Advantages of Cybermaterials

Because cybermaterials are created by a computer-aided design and fabrication process, the following advantages are expected.

1) Freeform materials and composites with optimized structures and functions can be fabricated.
2) Material design can be modified easily by revising the CAD program.
3) Materials to fit specific uses and environments can be produced.
4) Machining processes can be reduced, resulting in saved energy and resources, and environment conservation.
5) Flexible manufacturing, remote production, and just-in-time manufacturing of products is possible using the Internet.
6) Transportation and inventory of products can be reduced, resulting in saved energy and resources, and environment conservation.
7) New areas of cybermaterials science and engineering can be established.

2.3 Approach to Cybermaterials Engineering

Much development work is needed to establish cybermaterials engineering. It is necessary to provide CAD and CAM systems for materials design and processing, various databases and analytical tools, network systems, and applications. As the first steps to cybermaterials engineering, Miyamoto *et al.* are investigating the following two areas: the development of ceramic/

polymer photonic crystals; and the development of reactive rapid prototyping for freeform fabrication of intermetallics and composites.

2.3.1 Development of Ceramic/Polymer Photonic Crystals by Stereolithography

Photonic crystals can be created by forming three-dimensional periodic variations of dielectric constants [31, 32]. When electromagnetic waves with comparable wavelength to the periodicity of the crystal are propagated in every direction, the electromagnetic waves are totally reflected due to Bragg deflection.

This means that a bandgap of electromagnetic waves is formed similar to that of the electronic bandgap in semiconductors. In semiconductors, when electronic waves have comparable wavelengths to the periodicity of the crystal potential in each direction, the electronic waves are totally reflected by Bragg deflection. Thus electrons with these wavelengths cannot exist in the crystal, resulting in the formation of a bandgap. The bandgap of photonic crystals can be understood by an analogy of the bandgap in semiconductors. These artificial crystals can totally reflect light rays or microwaves without absorption.

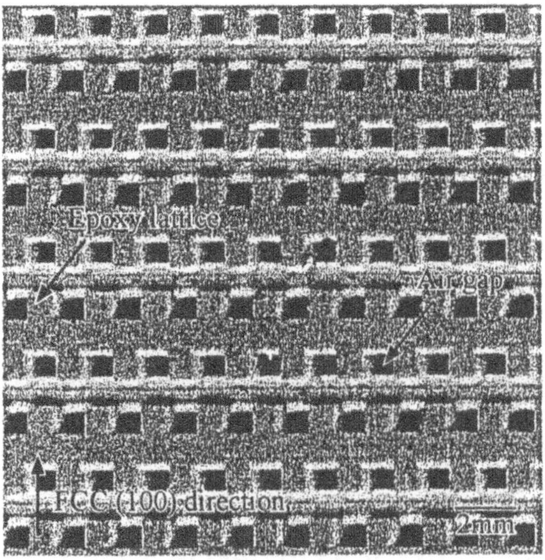

Figure 1.7. A three-dimensional lattice structure in epoxy photonic crystal with periodic variation in the dielectric constant fabricated by stereolithography.

Miyamoto *et al.* are designing photonic crystals composed of epoxy lattices of millimeter size, as shown in Figure 1.7. The periodic structures were processed by stereolithography, a rapid prototyping method [33]. To obtain the 3D structures, two-dimensional thin layers of epoxy resin photo-polymerized by UV-laser scanning are stacked automatically. Ceramic particles, such as SiO_2 and TiO_2, are mixed with photo-polymer epoxy resin to increase the dielectric constants of the electromagnetic media. The frequency range and the depth of the bandgap can be controlled by controlling the dielectric constant of a photonic crystal. Downsizing of photonic crystals can be achieved by increasing the dielectric constants, and this is a key factor for application to microwave devices. Figure 1.8 shows an example of the bandgap of a TiO_2/epoxy photonic crystal. Graded lattice spacings in a photonic crystal can be easily formed as well. Applications for these photonic crystals are expected in high-performance directional antennas for intelligent transportation systems of automobiles, in stratosphere platform telecommunication systems, and as barriers against electromagnetic interference.

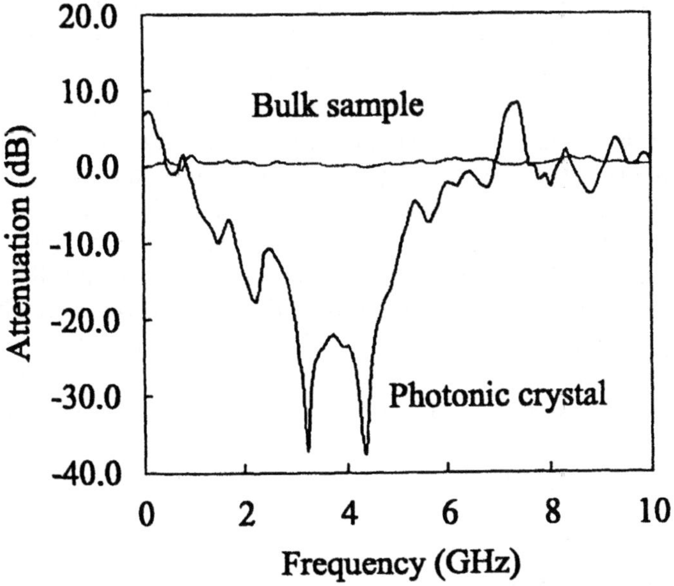

Figure 1.8. Transmission attenuation through TiO_2 dispersed epoxy photonic crystal and bulk sample.

2.3.2 Development of Reactive Rapid Prototyping for Freeform Fabrication of Intermetallic Materials and Composites

In the last few decades, many types of inter-metallic compounds, such as aluminides and silicides and their ceramic composites, have the potential for replacing conventional superalloys as advanced materials. Although these intermetallics have many excellent physical and chemical properties, they have extremely poor workability. Therefore, the method for producing these materials is generally limited to a near-net-shape process based on liquid casting or powder consolidation. Because the use of a mold is essential in these fabrication techniques, the product is limited to a simple shape. Consequently, for fabrication of intermetallic materials with complicated shapes that cannot be achieved by a conventional mold-based fabrication technique, a new moldless production method must be developed.

The concept of reactive rapid prototyping (RRP) is illustrated in Figure 1.9, using NiAl as an example [34]. When an aluminum droplet is injected onto a nickel powder bed, the two elements exothermically react and produce molten NiAl by self-propagating high temperature synthesis (SHS) [35], which immediately solidifies into a small bead. When the next droplet falls to a position close to the previous small NiAl bead, a new NiAl bead is similarly produced and is bonded to it. After a two-dimensional structure is produced on an x-y plane, nickel powder is added until it completely covers the structure on the former plane, and then a new structure is similarly formed by adding another nickel powder bed. The NiAl beads are bonded to each other in both horizontal and vertical directions.

Finally, a three-dimensional structure is produced after repeating the addition of the powder and the supply of molten aluminum droplets.

Miyamoto *et al.* are developing an RRP machine using a digitally controlled dispenser for aluminum droplets, an X-Y-Z stage controller for the powder bed, and a powder feeding port, which is schematically illustrated in Figure 1.10. They expect to be able to utilize this RRP to form turbine blades, gas nozzles, and molds of dense NiAl, which can be used in high-temperature, oxidizing and corrosive environments. This process can also be applied to other intermetallic systems such as Ti-Al, Co-Al, Fe-Al, Nb-Al and their composites. The RRP process can fabricate graded materials. For instance, it would be possible to form graded compositions of ceramics into the high-temperature section of a turbine blade using the digital fabrication processes.

2.4 Conclusions

Cybermaterials have the potential to reduce energy costs and resources, thereby minimizing their impact on the environment. These new-generation materials can also be designed with multiple functions. Graded structures and compositions can be easily formed using these digital fabrication methods. The graded lattice spacings in a photonic crystal or the graded composition of ceramics used for the high-temperature sections of a turbine blade could produce new high-performance devices. When these materials designs can be transmitted digitally using the Internet, physical long-distance transport of products can be reduced or even eliminated, leading to the new field of cybermaterials engineering.

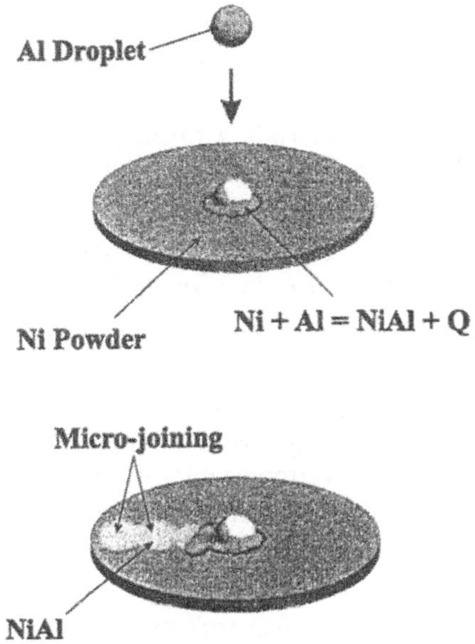

Figure 1.9. Fundamental concept of reactive rapid prototyping for NiAl.

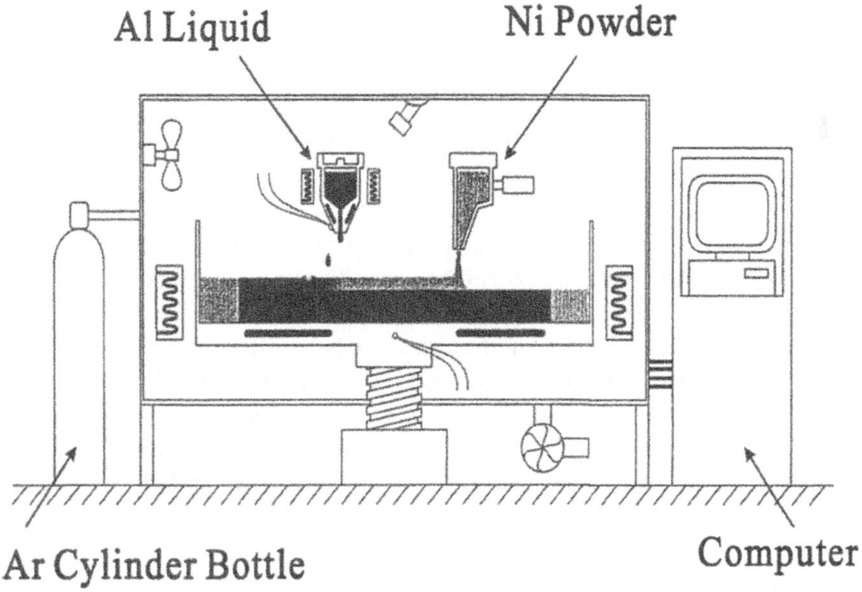

Figure 1.10. Schematic of the apparatus for reactive rapid prototyping.

3. ACTIVITIES OF THE FGM FORUM

3.1 Introduction

The concept of functionally graded materials (FGMs) was proposed in 1984 by materials scientists in the Sendai area of Japan as a means of preparing thermal barrier materials. Continuous changes in the composition, microstructure and porosity of these materials have resulted in gradients in such properties as mechanical strength and thermal conductivity. In 1987, a national project was initiated, entitled *Research on the Basic Technology for the Development of Functionally Graded Materials for Relaxation of Thermal Stress (FGM Part I)*, under the auspices of Japan's Agency of Science and Technology (STA).

At that time, many researchers and companies interested in FGM wanted to participate in this national government sponsored project. However, since funds were limited, it was impossible to accept all of the researchers' requests for participation. As a result, following the beginning of the FGM project, the STA established the FGM Forum in 1988 to allow greater participation.

Through this Forum, many researchers without government funding were able to join the FGM Project, enabling a greater amount of research on the development of FGMs to be conducted in Japan.

Since 1988, the FGM Forum has not only played an important role in the development of FGM domestically, but internationally as well. In the following section will be discussed the Forum's past activities, current state, and the expansion of the FGM Forum in the future.

3.2 History of Research and Development on FGM in Japan

Materials scientists in the Sendai area of Japan proposed the concept of functionally graded materials (FGMs) in 1984 as a means of preparing thermal barrier materials. Continuous changes in the composition, microstructure and porosity of these materials resulted in gradients in such properties as mechanical strength and thermal conductivity. In 1987, a government sponsored national project was initiated, entitled *Research on the Basic Technology for the Development of Functionally Gradient Materials for Relaxation of Thermal Stress* (April 1987 to March 1992, $11 Million, 30 Organizations)[36]. In 1992 when the first national project on FGM mentioned previously (FGM Part I) was finished, the results were samples of 300 mm square shell and 50 mm diameter hemispherical bowls for SiC-C FGM nose cones. The FGM concept is of interest not only in the practical design of super-refractory materials, but also in the development of various functional materials. Thus, in 1993, the second national project was initiated for research and development on FGMs as functional materials, entitled *Research on Energy Conversion Materials with Functionally Graded Structures* (FGM Part II) (April 1993 to March 1998, $9 Million, 21 organizations). This program aimed to apply functionally graded structure technology to improve energy conversion efficiency [37]. The project continued until fiscal year 1997.

Concurrently, the New Energy and Industrial Technology Development Organization (NEDO) sponsored a very important project on FGMs entitled *Precompetitive Processing and Characterization of Functionally Graded Materials* (April 1996 to March 2000, $2.5 Million, 4 Organizations).

The aim of this project was to develop large-scale metal-ceramic FGMs by the Spark Plasma Sintering (SPS) technique. This method was developed for fabricating FGMs with a gradient temperature. Many FGMs such as polyimide/Cu, PSZ/Ni and others have been successfully prepared using this technique [38].

However, the majority of these projects have been concerned with the grading of the mechanical properties of materials to reduce thermal stress, and not with the basic problem of functionally grading their physical and chemical properties. Basic research is essential for the creation of new functions using FGM technology. To fill this urgent need, a new national research program entitled, *Physics and Chemistry of FGMs* was initiated in 1996 with the financial support of the Ministry of Education, Science, Sports and Culture, MESSC (Grant-in-Aid for Scientific Research on Priority Areas; April 1996 to March 1999, $6.2 Million, 80 Organizations) [39]. A special feature of this basic research project is that it incorporated a wide range of fields, including physics, chemistry, biology, medicine, pharmacology, and agriculture. This project played an important role in promoting the investigation and application of FGMs, and many achievements have been made. These research activities are summarized in Table 1.3.

3.3 Early Activities of the FGM Forum

As mentioned above, the number of organizations that participated in the FGM Part I national project from academia and industry were limited. Therefore, the Functionally Gradient Material Forum (FGM Forum) was established in February 1988, to bring together many of the top scientists from universities, institutes and industry who were interested in FGM research and development, and who assumed an important role in that project without government funding.

Table 1.3. Progress of FGM research in Japan during the last 15 years.

1984	Original concept of FGM proposed by Dr Niino and other material scientists in Sendai area, Japan
1986	Feasibility study on 'The basic technology for the development of functionally gradient materials for relaxation of thermal-stress' under the auspice of Science and Technology Agency (STA)
1987-91	National project 'Research on the basic technology for the development of functionally gradient materials for relaxation of thermal-stress', FGM PART I
1988	Functionally Gradient Materials Forum (FGMF) 1st FGM Symposium organized by FGMF
1989	Germany-Japan FGM Seminar in Köln, Germany
1990	1st International Symposium on FGM in Sendai, Japan. International Advisory Committee of FGM (IACFGM)
1992	Feasibility study on R & D of FGMs as functional material under the auspice of STA. 2nd International Symposium on FGM and Japan, Germany, USA International Workshop in San Francisco, USA
1993-97	National project 'Research on energy conversion materials with functionally gradient structure FGM PART II
1994	The 3rd International Symposium on FGM in Lausanne, Switzerland
1995	Japan-Russia-Ukraine Workshop on Energy Conversion Materials (ENECOM-95) in Sendai
1996	The 4th International Symposium on FGM in Tsukuba Science City, Japan
1996-99	NEDO FGM project entitled ' Precompetitive Processing and Characterization of Functionally Gradient Materials'
1996-99	National research program entitled, "Physics and Chemistry of FGMs " MESSC
1998	The 5th International Symposium on FGM in Dresden, Germany
1999	The 2nd stage FGM Forum start NEDO International Symposium on FGM
2000	International FGM Workshop by JST in Tsukuba 6th International Symposium on FGM in Colorado, U.S.A.

The project started with scientists from 13 organizations. After the FGM Forum was established, new members from another four organizations also joined without government funding.

At the conclusion of the project, one of the new up-and-coming groups succeeded in producing a 50-mm diameter hemispherical ball composed of C/C composite using FGM coating technology that cleared the final target of this FGM Part I project, that is, usable in thermal environments at temperatures up to 2,000 K.

(a)

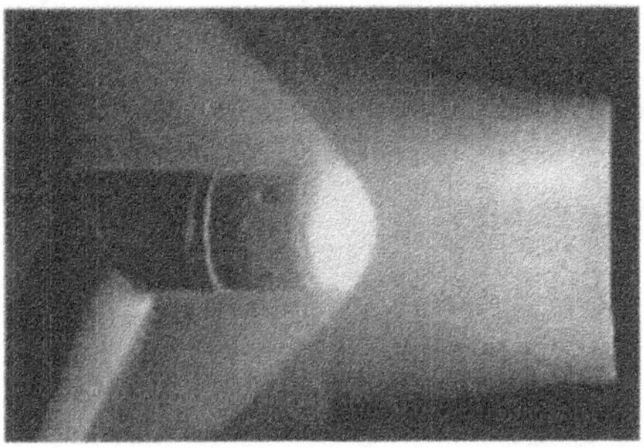

(b)

Figure 1.11. 50-mm diameter hemispherical C/C ball and the test using high temperature gas flow to evaluate hemispherical FGM.

This clearly confirms the pivotal role played by the FGM Forum. Figure 1.11 shows the 50-mm diameter hemispherical ball and the test using high-temperature gas flow to evaluate this hemispherical FGM.

3.4 Other Vital Roles of the FGM Forum

In addition to its primary objectives, the FGM Forum has other very important activities which include: publicizing the activities of national projects, gathering information on FGMs worldwide and surveying possible applications for FGMs as structural and functional materials. The forum continues to hold symposiums and workshops and to publish bulletins. Furthermore, the FGM Forum has continued to carry out various surveys in a search for outstanding research and development topics for future FGM-related projects.

3.4.1 Organization of the FGM Forum

To achieve its objectives, the FGM Forum has the organizational format shown in Figure 1.12.

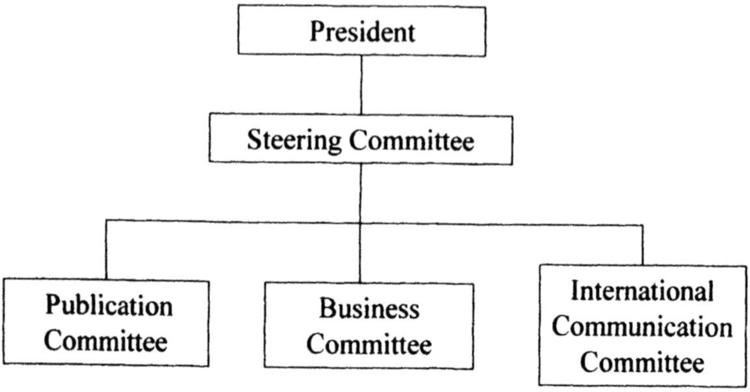

Figure 1.12. Organization of FGM Forum.

3.4.2 Committee Activities

The **Publication Committee** publishes the *FGM News* three to four times per year, including one full English edition. In the Japanese edition, titles, abstracts, figures and table captions translated in English are attached.

The **Business Committee** holds an annual domestic symposium, and workshops several times a year. The papers submitted to the domestic symposium are refereed, and are published in the proceedings thereafter. Another responsibility of the Business Committee is to find new business for the FGM Forum, including technical transfer of FGM technologies invented by universities and institutes to industry.

The **International Communication Committee**: The first FGM International Symposium organized by the FGM Forum was held in Sendai in 1990. It was so attractive that about 400 participants including scientists from abroad participated. During the Symposium, an International Advisory Committee was established led by Professor Mitsue Koizumi, who was then vice President of the FGM Forum (now current President of the FGM Forum). In the first meeting of IACFGM held in Sendai, the IACFGM decided to hold the next International Symposium in the US two years later. Since then, every two years, International Symposiums have been held, in San Francisco, Lausanne, Tsukuba and Dresden. The next one is to be held in Colorado. At every Symposium, the IACFGM, whose office is located in the FGM Forum, cooperated with the organizing committee, and the International Communication Committee of the FGM Forum in Japan assisted in the activities of the IACFGM.

Table 1.4. Survey activities on the FGM Forum.

Year	Title of the survey	Results
1987	Survey of the Multi-elements composed technology of the material	Promote FGM Part I
1989	FGM survey in Europe	
1990	Survey for the design standard of FGM	
1991	Needs survey of FGM	Promote FGM Part II
1993	Feasibility Studies of FGM to Future Transportation Systems	NAL-NASDA Cooperative Research Program
1995	The possibility survey of electric power generation from thermal energy	
1999 | 2000	Survey for the new projects	Aim to Establish New R&D National Project??

The **Steering Committee** manages all other committees' budgets and each committee's plans, and makes a continual effort to develop new discussions through the use of surveys. The Steering Committee organizes temporary committees to survey new topics for discussion. The survey activities of the FGM Forum are listed in Table 1.4. The most interesting survey, the *Needs Survey of FGM* in 1991, led to a new national R&D project (FGM Part II). Another interesting survey, *Feasibility Studies on FGM for Future Transportation Systems*, in 1993 was supported by NASDA. As a result of that survey, a NAL-NASDA cooperative research program started. Moreover, these surveys were instrumental in generating the national research project entitled, *Physics and Chemistry of FGMs* MESSC (1996-1999).

3.4.3 Current Problems of the FGM Forum

Forum members had included representatives from 63 companies and 31 individuals at the forum's inception, but now in March 2000, the numbers now stand at 30 companies and 95 individuals. The transition in the numbers of members is shown in Figure 1.13.

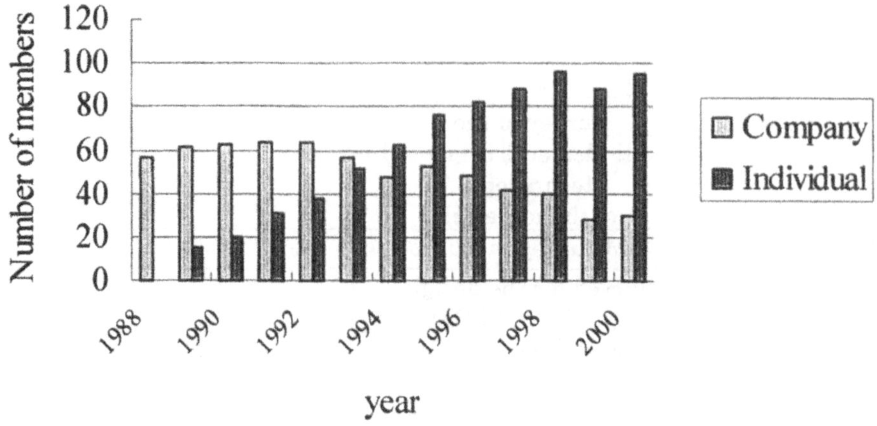

Figure 1.13. Transition of FGM members.

The number of corporate members gradually began to decrease after the completion of FGM Part I. Many corporate members that had participated from the beginning of the FGM Forum had had an interest in only the structural materials and they considered that it was impossible to make use of information about the structural FGM materials from the FGM Part II project on thermoelectric and thermion energy conversion technologies. As a result, they left the FGM Forum after the completion of the FGM Part I project. This trend was accelerated by the overall economic unrest in the aftermath of the collapse of the bubble economy in Japan.

These circumstances lead the FGM Forum management to believe that most corporate members did not have an interest in FGM itself, but in applications of FGM.

Nonetheless, it is a fact that corporate memberships have decreased remarkably and conversely, individual memberships have increased substantially. Thus the FGM Forum management has to plan to recruit corporate members, while operating with a focus on its individual members.

3.5 The Second Stage of the FGM Forum

The FGM Forum thus had to be reformed to correspond to these changes in the membership environment. The fundamental policy of the Second Stage of the FGM Forum as of 1999 is as follows:

1) Improve efficiency
2) Lower costs
3) Focus on individual members
4) Increase the membership (especially international members)

5) Promote technology transfer from universities and institutes to industry

To achieve these policies, the FGM Forum is going to do following items:

1) Upgrade the FGM Internet Website
2) Construct an FGM intranet
3) Upgrade the FGM Database

On a practical basis, they are going to improve the FGM Official Website and construct an intranet for FGM which only FGM Forum members can access. Through this intranet, FGM Forum members will be able to access the following information:

1) FGM Database
2) FGM News
3) Project Forum

After the intranet for the FGM Forum is constructed, international members from abroad will be able to easily access the FGM Database, FGM News and Project Forum and obtain useful information about FGM.
They hope that many researchers who read these proceedings will join the FGM Forum and that activities related to FGM continue to expand not only in Japan but also throughout the world.

REFERENCES

1. Kawasaki, A. and Watanabe R., J. Japan Inst. Metals, 51 (1987), 525.
2. Niino, M., Hirai, T. and Watanabe, R., J. Japan Soc. Comp. Mater., 13 (1987), 257.
3. Nishida, I., FGM News, 21 (1993), 17.
4. Hirai, T., Mater. Sci. and Tech., Eds. R.W. Cahn *et al.*, VCH Verlag, 17B (1996), 293.
5. Hirai, T., Research Report 1998, Grant-in-Aid for Scientific Research on Priority Area, No. 274, The Ministry of Education, Science, Sports and Culture of Japan, (1998).
6. Kitaguchi, S., Ceramics (Bulletin of the Ceramics Society of Japan), 29 (1994), 182.
7. Hirai, T. and Sasaki, M., Ceram. Int., 17 (1991), 275.
8. Shoda, Y., Kude, Y., Uemura, S., Saitoh, T., Wakamatsu, M. and Niino, Y., Ceramic Transactions, Vol. 34, J.B. Holt *et al.* (Ed.), (1993) Amer. Cer. Soc., 125.
9. Kawasaki, A. and Watanabe, R., J. Japan Inst. Metals, 56 (1992), 472.
10. Kawasaki, A., Chien-Hung, Y. and Watanabe, R., J. Japan Soc. Powder and Powder Metall., 43 (1996), 295.
11. Erdogan, F. and Chen, T. C., Materials Science Forum, 308-311 (1999), 917.
12. Hofinger, I., Bahr, H.A., Balke, HG., Hsler, Kirchhoff, C. and Weiss, H.J., Materials Science Forum, 308-311 (1999), 19.
13. Tsuda, K., Ikegaya, A., Nomura, T., Isobe, K., Kitagawa, N., Nakado, M. and Arimoto, H., Sumitomo Denki (Technical Report of Sumitomo Electric Ind.,) No. 147 (1995), 71.
14. Hirai, S., Noguchi, H., Saka, T., Kanno, T. and Kuwabara, M., Materia Japan (Bulletin of Japan Inst. Metals), 37 (1998), 510.

15. Li, J. F., Nishio, N., Kawasaki, A. and Watanabe, R., J. Japan Soc. Powder and Powder Metall. 41 (1994), 653-657..
16. Hirano, T., Whitlow, L. W. and Miyajima, M.: Ceramic Transactions Vol. 34, edited by Holt, J.B., *et al.* (Eds.), Amer. Cer. Soc., (1993), 23.
17. Wang, X., Masumoto, H., Someno, Y. and Hirai, T., J. Japan Inst. Metals, 62 (1998), 1069.
18. Wang, X., Masumoto, H., Someno, Y. and Hirai, T., J. Japan Soc. Powder and Powder Met., 46 (1999), 180.
19. Kawasaki, A. and Watanabe, R., J. Japan Soc. of Powder and Powder Metall., 37 (1990), 287.
20. Ono, M., Li, J. F. and Watanabe, R., Collected Abstracts of Spring Meeting of Japan Inst. Metals, (2000), 398.
21. Omori, M., Kakita, T., Okubo, A. and Hirai, T., J. Japan Inst. Metals, 62 (1998), 986.
22. Watanabe, R., Ichiki, K., Hayama, T. and Kawasaki, A., Materials Science Forum, 308-311 (1999), 19.
23. Kuroki, Y., Kawasaki, A. and Watanabe, R., J. Japan Soc. Powder and Powder Metall., 38 (1991), 700-706.
24. Watanabe, R. and Miyajima, M., Proc. 7th FGM Symposium, [FGM'94], Tokyo, November 1994, FGM Forum, Tokyo, (1995), 139-142.
25. Fukumoto, M., Ito, M., Tanaka, Y., Yakabe, H. and Kikuchi, K., J. Japan Inst. Metals, 62 (1998), 449.
26. Ohyanagi, M., Tsujikami, T., Sugahara, S., Koizumi, M. and Levashov, E.A., Materials Science Forum, 308-311 (1999), 145.
27. Zhao, G., Kouzuka, H., Lin, H., Takahashi, M. and Yoko, T., Thin Solid Films, 340 (1999), 125.
28. Miyamoto, Y., Kaysser, W.A., Rabin, B.H., Kawasaki, A. and Ford, R.G., *Functionally Graded Materials: Design, Processing, and Applications*, Kluwer Academic Publishers, (1999).
29. Bever, M.B.and Duwez, P.E., Gradients in Composite Materials, Mater. Sci. Eng., 10 (1972), 1–8.
30. Griffin, C., Daufenbach, J. and McMillin, S., Desktop Manufacturing, LOM vs. Pressing, Am. Ceram. Soc. Bull., 73-173 (1994),109–103.
31. Joannopoulos, J.D. Maede,R.D. and Winn J.N. , *Photonic Crystals, Molding the Flow of light*, Princeton University Press, (1995).
32. Yablonovitcii, E., *Photonic Crystals*, J. Modern Optics, 41 (1994), 173-194.
33. Kirihara, S., Miyamoto, Y. and Kajiyama, K., *Fabrication of Ceramic/Expoxy Photonic Crystals by Stereolithography*, Ceramic Engineering & Science Proceedings, Am. Ceram. Soc., to be published.
34. Matsuura, K., Kudoh, K., Kirihara ,S. and Miyamoto, Y., *Solid Freeform Fabrication of Intermetallics and Their Ceramic Composites by Reactive Rapid Prototyping*, ibid.
35. Rogavhev, A. S., Varma, A. and Merzhonov, A G., *The Mechanism of Self-Propagating High Temperature Synthesis of Nickel Aluminides, Part I; Formation of the Product Microstructure in a Combustion Wave*, Int'l. J. of SHS, 2 (1993), 25-38.
36. Niino, M. and Maeda, S., ISIJ International, 30 (1990), 699.
37. Miyamoto, Y. Niino, M. and Koizumi, M., Proc. PacRim Meeting at Austria, (1996).
38. Ohmori, M., Kawahara, M., Sakai, H., Okubo, A. and Hirai, T., J. Jpn. SK. Powd. & Powd. Metall., 41 (1949), 649.
39. Hirai, T., *Physics and Chemistry of Functionally Graded Materials*, Research Report 1998, MESSC (1998).

Chapter 2

MODELING AND SIMULATION

1. COMPUTER-AIDED DESIGN OF GRADED POWDER COMPACTS

1.1 Introduction

Pressureless sintering can be a cost-effective process for producing large-scale parts of functionally graded materials (FGMs), but variations in sintering shrinkage among graded layers often cause sintering defects, such as distortion and cracks [1, 2]. Successful sintering of graded powder compacts requires a uniform shrinkage curve for every layer, which can be attained by for instance by selecting the particle size or adding some element [3]. Optimization of the graded structure, e.g. the number and the thickness of layers, can be also effective because adjusting the sintering balance is not always easily accomplished. Although several attempts have been reported, as reviewed in Refs.[4, 5], a trial-and-error approach is still inevitable. When the sintering curves differ among the laminated layers, internal stress is generated by constraint and affects the shrinkage of each layer. It is difficult to predict the effects of constraint among the layers on the densification only from the shrinkage curves because the magnitude of constraint results from combinations of shrinkage force and flow resistance in each layer, and these do not appear on shrinkage curves.

Recently, methods for analyzing the sintering process of laminated or graded powder compacts have been developed for the purpose of reducing experimental iterations [6, 7]. Shrinkage force and flow resistance were taken into account as the sintering stress and viscosity, respectively, in a constitutive model for sintering. In this section is described a scheme for controlling sintering defects in FGMs which is demonstrated on a computer using the finite element method. Step-wise graded structures consisting of metallic/ceramic powder mixtures were designed as a model FGM. Densification, distortion and stress in the graded powder compacts during sintering were computed for some designed compacts to find a condition without warping, high internal stress and insufficient densification.

1.2 Design Scheme Using Simulation

1.2.1 Analytical Method of Sintering Process

Sintering curves can be characterized by three parameters; the shrinkage start temperature, the shrinkage rate and the total shrinkage[3]. In numerical simulations, however, the sintering behavior is expressed by a constitutive model. In this Section, a constitutive model developed in a previous study[8] is used;

$$\dot{\varepsilon}_{ij} = \frac{1}{2\eta_{MC}}\frac{1}{\rho^{2n-1}}\left\{\sigma'_{ij} + \delta_{ij}\frac{2}{9f^2}\left(\sigma_m + \sigma_{SMC}\right)\right\} \tag{2.1}$$

where $\dot{\varepsilon}_{ij}$ is the strain rate ($i, j=x, y, z$), ρ is the relative density, σ'_{ij} is the deviatoric stress, δ_{ij} is the Kronecker symbol, σ_m is the hydrostatic stress, n and f are parameters representing the porous structure, η_{MC} and σ_{SMC} are the viscosity and the sintering stress of the powder mixtures, respectively. The viscosity is dependent on temperature and grain size. The sintering stress is a function of the size of the powder particles and relative density.

Based on the constitutive model, finite element analysis can be performed. The nodal forces for each element are given by [6]

$$\{P\} = \int_V [B]^T [D][B] dV \{u\} + \int_V [B]^T \{S\} dV \tag{2.2}$$

where [B] is the strain-rate matrix, [D] is the matrix correlating the stress components with the strain-rate components, {u} is the nodal velocity vector and {S} is the sintering stress vector. Since the needed conditions are high density without crack formation, as well as geometrical requirements, densification and internal stress in the compacts should be checked. The stress generated in a sintering body is given by

$$\{\sigma\} = [D][B]\{u\} + \{S\} \tag{2.3}$$

1.2.2 Overall Flow of CAD

Figure 2.1 shows the positioning of the sintering simulation in computer-aided design (CAD) of FGMs. At first, according to a required material combination, a tentative plan of the compositional profile and the conditions of the powder process can be set up. An analysis of the powder forming is the first significant step. For example, graded powder compacts may warp even during cold isostatic pressing due to variations of powder characteristics in the layers. A density distribution obtained from an analysis of the powder forming can be utilized as the initial state of the sintering analysis, which is important for predicting distortions in sintered bodies. A database for sintering analysis should consist of parameter values in constitutive models, which are related to the sintering stress and the viscosity with relative density, pore geometry, grain growth and so on. As results of the sintering analysis, the distributions of the relative density and grain size in sintered bodies can be obtained as well as stress and strain. If criteria for cracking, splitting or delamination during sintering have been established, they would complete the sintering analysis. The thermal stress analysis can be conducted separately after the sintering analysis or simultaneously using the visco-elastic finite element method.

1.3 Sintering Simulation

1.3.1 Subject

A metallic/ceramic FGM consisting of six layers was adopted as a subject in this study. As components of the FGM, six different powders, F1 to F6, of stainless steel (mean size 6μm) and alumina (mean size 0.25μm) were prepared with mixing ratios of 100/0, 80/20, 60/40, 40/60, 20/80, 0/100, respectively. Figure 2.2 shows the sintering behavior of the six powder compacts. The characteristics of this system are summarized as follows [8]:

1) Sintering of F1 (100% steel) starts at a lower temperature than the others do,

2) The total shrinkage of F6 (100% alumina) is much higher than that of the others, and

3) The final relative densities of F3 and F4 are lower.

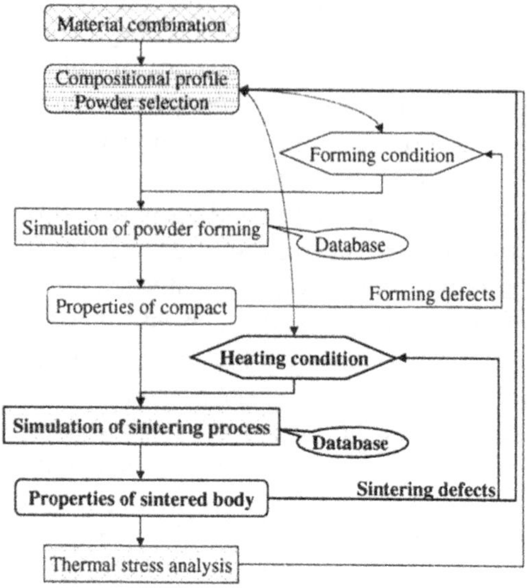

Figure 2.1. CAD for FGMs fabricated by sintering.

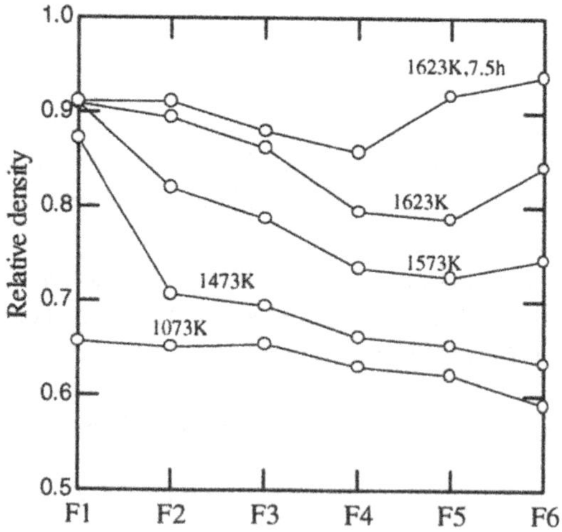

Figure 2.2. Densification behavior of each layer.

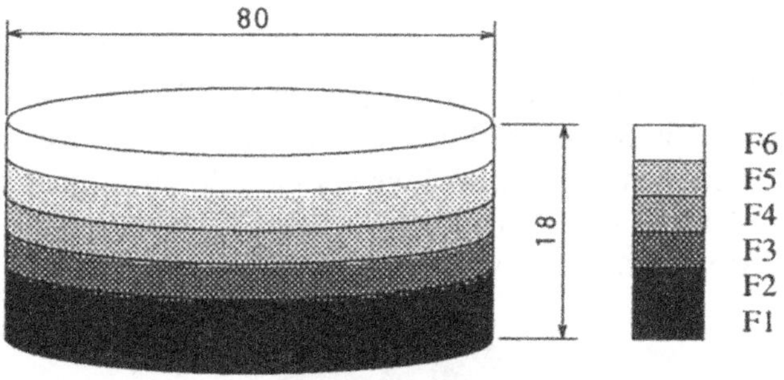

Figure 2.3. Geometry of graded powder compacts.

1.3.2 Design

Figure 2.3 shows the graded powder compacts in the shape of a disk prepared for analysis. The means of controlling sintering defects examined in the present study are:

i) gaining the thickness of a certain layer as a core,

ii) adjusting the initial relative density to balance the total shrinkage,

iii) selecting the particle size to modify the shrinkage rate.

According to the above methodology, the graded powder compacts MC1-3 were designed as listed in Table 2.1. Figure 2.4 shows heating process and the shrinkage-time curve of each layer for the designed FGM.

Table 2.1. Design for FGM.

MC0	Basis (without improvement)
MC1	F1 layer with 4 times thick as the others
MC2	Initial relative density of 0.64 for all layers
MC3	Twice the size of powder particles for F6 in MC0
MC4	1.8 times the size of powder particles for F6 in MC2

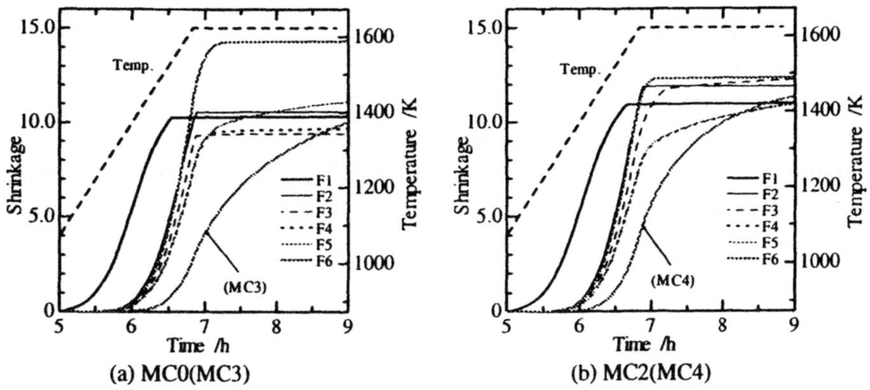

Figure 2.4. Shrinkage behavior of each layer.

1.3.3 Simulation

Figure 2.5 shows the distorted grid patterns of MC0 and MC1. From the calculated results, it is predicted that warping of the disk occurs due to the lack of sintering balance of the six layers, especially the higher shrinkage of the F6 layer. Obtaining the thickness of the F1 layer is not effective for controlling warping for the present system as MC1 indicates.

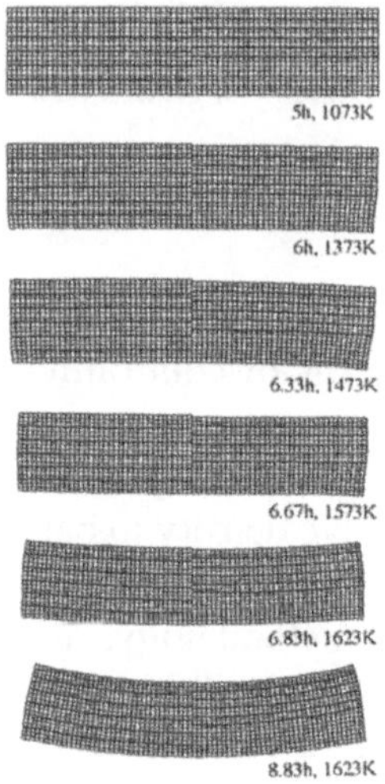

(a) MC0 (b) MC1

Figure 2.5. Warping of graded powder compacts during sintering. (Top:F6, Bottom: F1)

The deformation behavior and the distributions of hydrostatic stress induced during sintering are shown in Figure 2.6. Although warping was not suppressed much in MC2, MC3 became flat at the end of sintering. The increase in the tensile stress in MC3, however, may imply increased probability of cracking. Furthermore, as can be seen in Figure 2.7, the decrease in the finial relative density of the F6 layer for MC3 may not be desirable in view of durability.

Another graded compact, MC4 (see Table 2.1), can be designed to improve the above-mentioned weak points of MC2 and MC3. As shown in Figures 2.6(d) and 2.7(d), the sintered MC4 is flat and both the internal stress and relative density are moderate.

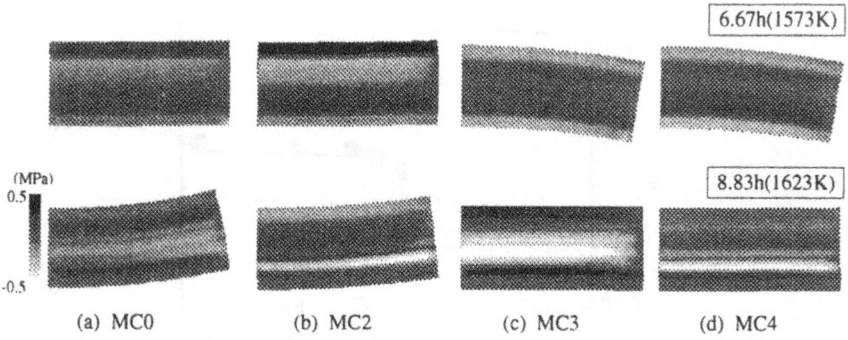

Figure 2.6. Distributions of hydrostatic stress in graded powder compacts during sintering.

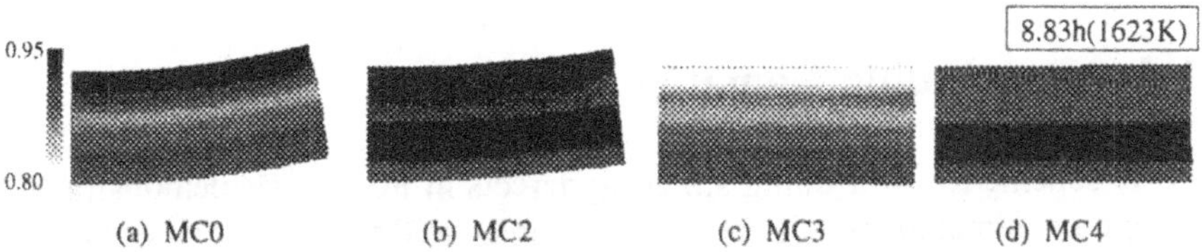

Figure 2.7. Distributions of relative density in sintered FGMs.

1.4 Practical Fabrication of FGMs

Sintering defects were controlled in the experiment, but only by means of selecting the particle size (iii) in Section 1.3.2). First, an unimproved graded powder compact with the size of $\phi 38.5 \times 7.74$ was sintered (MC0e). Figure 2.8 shows the change in geometry of the disks during sintering. Warping actually occurred as predicted in Section 1.3.3, while the F1 layer was separated along with F2 and many cracks formed in the F6 layer[9]. Hence a layer made of alumina powder with a mean size of 0.37 μm was replaced as F6, but stainless steel powder with a mean size of 70 to 80 μm was also employed specially for F1 to avoid separation (MC3e). As can be seen in Figure 2.8, flatness was achieved in MC3e, and with no delamination, though cracks in F6 still appeared. Although damage during sintering was not taken into account for computation at present, the anticipated suppression of warping by selecting the particle size was successful.

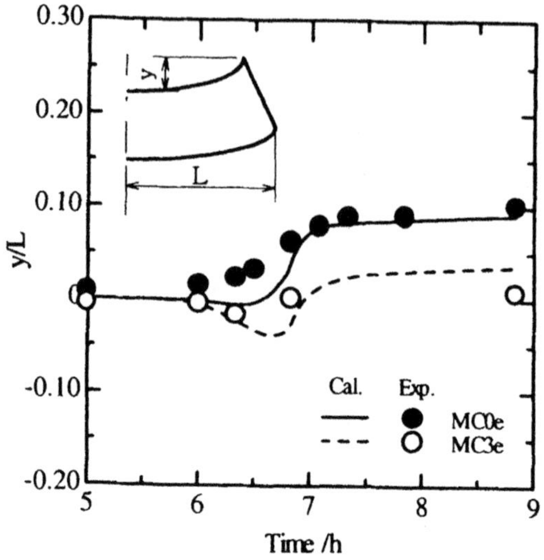

Figure 2.8. Control of warping by powder size selection.

1.5 Concluding Remarks

A scheme for controlling sintering defects in FGMs was demonstrated in computer simulations. A metallic/ceramic FGM without warping was designed using sintering simulation and actually fabricated.

Predicting sintering damage is required in CAD formulations of FGMs to make it practicable. Since preparation of the required powder compacts is also essential, material and process parameters should be stored as a database available during simulations.

2. TEMPERATURE AND STRESS DISTRIBUTIONS IN FGM PLATES UNDER THERMAL SHOCK

2.1 Introduction

Functionally graded materials (FGMs) have been attracting considerable attention in extremely high-temperature structural ceramic applications. These materials are composed of different components such as ceramics and metals which have a continuous or gradual variation of the mixing rate across the thickness. Such a microstructure is expected to reduce thermal stresses and to improve the thermal shock properties of the materials.

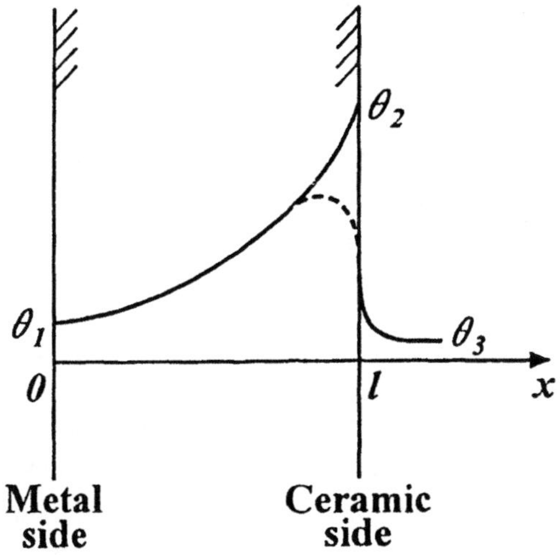

Figure 2.9. Temperature distributions in FGM plate under thermal shock.

The objective of the present study is to present a numerical technique for analyzing one-dimensional transient temperature distributions in a ceramic-metal-based FGM plate, in relation to both the temperature-dependent thermal properties and the gradual variation of the thermal properties of the plate [10-12]. The FGM plate shown in Figure 2.9 is assumed to be initially in a steady-state temperature gradient, as it is exposed to high temperature on the ceramic surface and to low temperature on the metallic surface, as associated with its in-service performance. The ceramic surface of the FGM plate is then rapidly cooled using a cold medium. The transient temperature in the FGM plate was analyzed numerically as a model for an alumina-nickel FGM system. Compared with other methods[13,14], the numerical technique proposed here for analyzing the temperature distributions is quite simple and widely applicable for various boundary conditions of FGMs.

2.2 Analysis

The analytical technique has been presented elsewhere [10,11], and will be briefly explained below. The unsteady-state normalized equation of the heat transfer is expressed as

$$(\rho C)^* \frac{\partial T}{\partial \tau} = \frac{\partial}{\partial \xi}(\lambda^* \frac{\partial T}{\partial \xi}) \tag{2.4}$$

with the initial condition

$$T = T_1 \quad at \quad \tau = 0 \tag{2.5}$$

and boundary conditions

$$\begin{aligned} &T = 0 \quad on \quad \xi = 0 \\ &\frac{\partial T}{\partial \xi} = \frac{\beta}{\lambda^*} T_s \quad on \quad \xi = 1 \end{aligned} \tag{2.6}$$

where the following non-dimensional variables and thermal properties are used.

$$T = \frac{\theta - \theta_3}{\theta_2 - \theta_3}, \quad \tau = \frac{\kappa_{ci} t}{l^2}, \quad (\rho C)^* = \frac{\rho C}{(\rho C)_{ci}}, \quad \beta = \frac{hl}{\lambda_{ci}}, \quad \xi = \frac{x}{l} \tag{2.7}$$

T represents non-dimensional transient temperature, T_I the initial temperature distribution in the plate, θ the actual temperature shown in Figure 2.9, τ the Fourier number, t the actual time, l the thickness of the plate, κ the thermal diffusivity, λ the thermal conductivity, ρC the heat capacity per unit volume (ρ: density, C: specific heat), β the Biot number, h the heat transfer coefficient on the ceramic surface, ξ the non-dimensional coordinates and T_s the temperature on the ceramic surface. Hereafter, the material properties with superscript * indicate non-dimensional values and properties with subscript ci indicate the material properties of the ceramics at room temperature.

The following variable transformation from T to G is used here to reduce Equation 2.4 to a simple differential equation [15].

$$G = \int_0^T \lambda^*(T,\xi) dT \tag{2.8}$$

Using Crank-Nicolson's differential expression for Equation 2.8, a finite differential equation is derived:

$$\begin{aligned} RG_{m+1,n+1} - (2R+1)G_{m,n+1} + RG_{m-1,n+1} &= \\ - RG_{m+1,n} + (2R-1)G_{m,n} - RG_{m-1,n} \end{aligned} \tag{2.9}$$

where the following notations for the finite variables are used.

Time: $\tau = n\Delta\tau$, (n = 0,1,...,N)

Distance: $\xi = m\Delta\xi$, (m = 0,1,....,M)

Temperature: $G(\xi, \tau) = G_{m,n}$

Also, $R = \Delta\tau\kappa^*/2\Delta\xi^2$. The values of $G_{m,n}$ are then calculated numerically and inverse-transferred to $T_{m,n}$.

Referring to the technique proposed by Tanigawa *et al.* [13,14], the thermal stresses in the FGM plate were analyzed, where the non-dimensional stress is defined as

$$\sigma_y^* = \frac{\sigma_y}{E_{\bar{a}}\alpha_{\bar{a}}(\theta_2 - \theta_3)}, \tag{2.10}$$

σ_y is the thermal stress, E the Young's modulus and α the thermal expansion coefficient.

In this section, Awaji *et al.* considered a model of a ceramic-metal-based FGM in which alumina was selected as the ceramic and nickel as the metal. The thermo-mechanical properties of these materials are shown in Ref.[10]. The local volume fraction of the metal in the FGM plate, $f_m(\xi)$, is described by

$$f_m(\xi) = 1 - \xi^p \tag{2.11}$$

where p is the parameter of volume fraction. To keep things simple, the local values of the thermo-mechanical properties of the FGM were postulated using an elementary law of the mixture model.

2.3 Results and Discussion

The non-dimensional temperature and stress distributions in the FGM plate were analyzed. The plate initially had a steady-state temperature gradient, and was then rapidly cooled on the $\xi = 1$ surface using a cold medium. Sintered FGM materials generally have residual stresses at room temperature because of the differences in thermal expansion between the constituent materials. In this section, however, Awaji *et al.* ignored the residual stresses that are expected to be initially distributed in the FGM plate.

Figure 2.10 shows the transient temperature distributions in the alumina-nickel FGM plate with $p = 1$, where the ceramic surface was cooled with cold water with $h = 4{,}600$ $Wm^{-2}K^{-1}$. The dotted curve indicates the analyzed temperature distributions without taking the temperature dependence of the thermal properties of materials into account. The solid curves are the analyzed temperature distributions with the temperature dependence taken into account. The Fourier number is varied from $\tau = 0$ to 0.1 in 0.02 steps. It was clear that the temperature dependence of the thermal properties induced marked effects on the temperature distributions in the FGM plate.

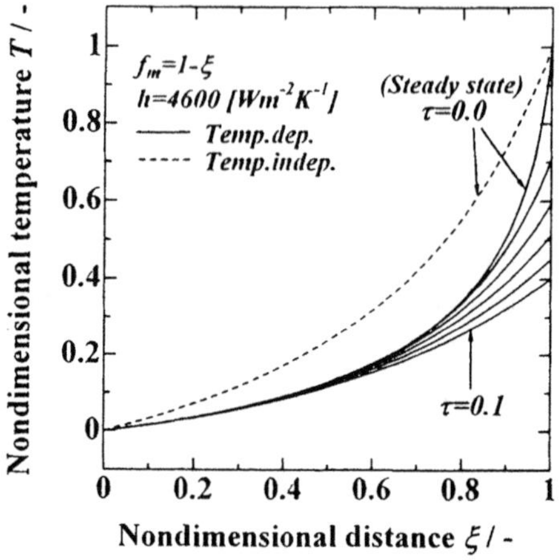

Figure 2.10. Temperature distributions in FGM plate with volume fraction, $f_m = 1-\xi$, taking temperature dependent properties into account.

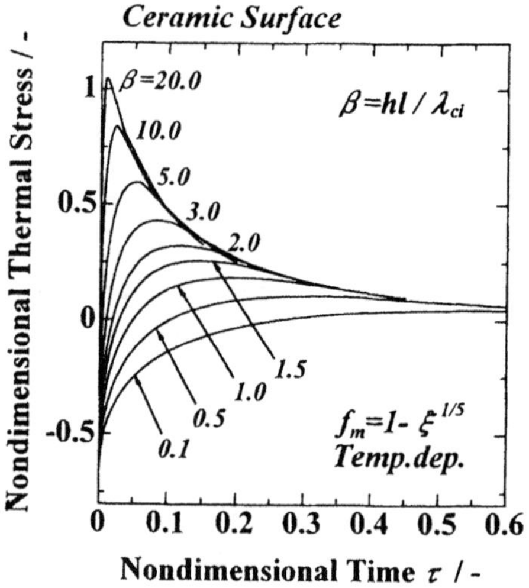

Figure 2.11. Relationship between thermal stress on ceramic surface and time with various Biot numbers, β.

Figure 2.11 shows the relationship between the stress distributions on the ceramic surface and the non-dimensional time with various Biot numbers, β. These curves are quite close to the widely acknowledged variations of maximum transient thermal stresses in a cooled homogeneous plate, but it should be noted that the initial stresses on the ceramic surface are negative in this figure.

2.4 Conclusions

Awaji *et al.* presented a numerical technique for analyzing one-dimensional steady-state and transient temperature distributions in a ceramic-metal-based FGM plate, in relation to both the temperature-dependent thermal properties and gradual variation of the thermal properties of the FGM. Their results indicated that these temperature-dependent thermal properties induced marked effects on transient temperature distributions in the FGM plate, and also that the temperature distributions in the FGM plate with gradual variation of the volume fraction of metals were analyzed successfully using our proposed simple variable transformation technique. Compared with methods proposed by other researchers, their technique is quite simple and widely applicable.

The transient thermal stress distributions in the FGM plate were also analyzed referring to the technique proposed by Tanigawa *et al.*, relative to both the temperature-dependence and gradual variation of the thermo-mechanical properties of the FGM.

3. FINITE ELEMENT ANALYSIS FOR EVALUATION OF RESIDUAL STRESS

3.1 Introduction

Functionally graded materials (FGMs) are remarkable heat-resistant materials for relaxation of thermal stress. Recently, gradually dispersed diamond composite materials have been fabricated using the Self-propagating High-temperature Synthesis (SHS) method [16]. However, in a system with two layers composed of a diamond dispersion layer and a matrix layer, high residual stress will likely occur near the interface of the layers. Therefore, gradual dispersion of the diamond particles in the matrix reaction is required to reduce the residual stress. A functionally graded material is a laminate material with smooth mechanical properties that avoid stress mismatch. By considering laminate structures, the kind of material and its thickness, mismatches of residual stress can be reduced. The objective of this study is to develop an evaluation system for macroscopic residual stress using finite element analysis and to calculate the residual stress in the gradually dispersed diamond material (Diamond/TiB_2/Si) fabricated by the SHS method.

For an accurate evaluation, it is essential to use a detailed model of the FGM. Therefore, a finite element modeling system with image processing (Image Mesh Generation System) was developed to make detail models for the material design of functionally graded materials [17]. The shape and material composition are automatically recognized from an image of the material, and the finite element model can be easily generated. As an example of using the developed mesh generation system for numerical analysis, Tsujikami *et al.* tried to calculate the internal microscopic residual strain for the WC/Ni fabricated by the SHS method. The calculated results were compared with the results by neutron diffraction measurement.

3.2 Macroscopic Residual Stress

3.2.1 Model of Gradually Dispersed Diamond Material

In this simulation, Tsujikami *et al.* evaluated the steady-state thermal residual stress of the number of graded laminas by FEM, and investigated the effect of the graded layer. Axis-symmetric cylindrical specimens with Diamond/TiB_2/Si graded material were utilized and this allowed a two-dimensional model to be employed. Figure 2.12 shows the scheme for an axis-symmetric cylindrical model of gradually dispersed diamond material. The specimen's dimensions are 5 mm height and 8 mm radius. Each specimen has two graded region thicknesses: one is 3.33 mm and the other is 1.67 mm, as shown in Figure 2.13. These specimens consisted of an upper diamond layer; a Diamond/(TiB_2/Si) graded middle layer and a bottom layer; (TiB_2/Si). The volume fractions of the Diamond and TiB_2/Si in the upper layer were 70% and 30%, respectively. The volume fractions of the TiB_2 and Si in the bottom layer were 50%. The mechanical properties of each layer were decided by rule-of-mixtures. And the volume fractions for the mechanical properties in the middle layer were decided using the equation $V_{fTiB2}=(x/t)^P$, where P is the power law exponent. In the case of $P=1$, they employed some models in which the middle layer was divided into 0, 1, 2, 3, 4, 5, 10 and 20 laminas. And a model with same number of finite elements was employed to ignore the effect of element division. The number of elements was 9600 and the number of nodes was 4941. Thermal residual stress for these models was investigated for cool down from the manufacturing temperature (2275 K) to room temperature (295 K). Table 2.2 shows the mechanical properties of the Diamond, TiB_2 and Si layers.

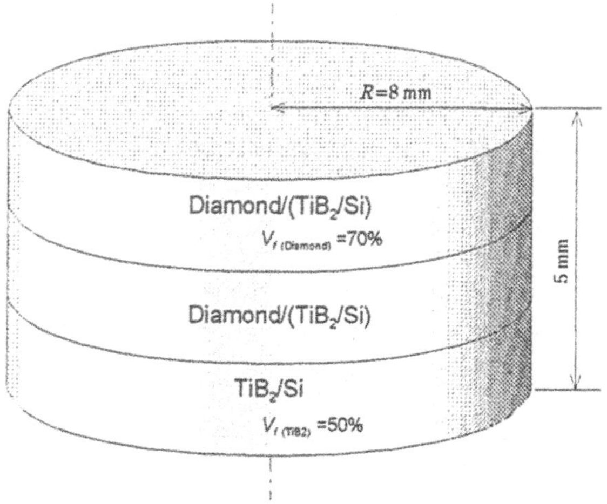

Figure 2.12. Axis-symmetric cylindrical model of three-layered FGM.

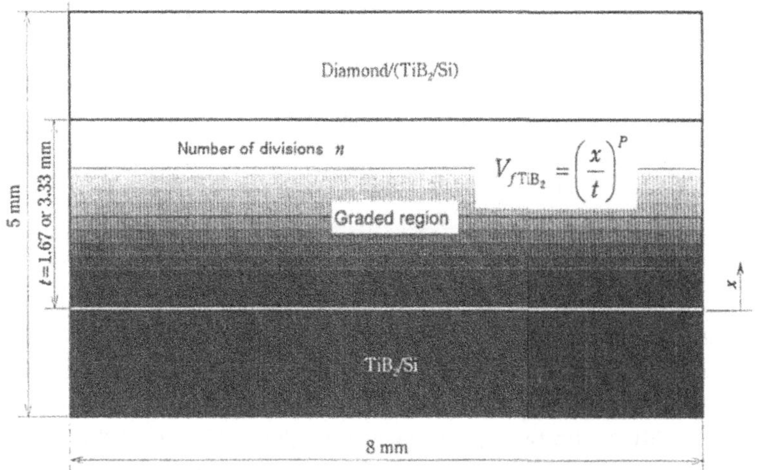

Figure 2.13. Two-dimensional model of three-layered FGM.

Table 2.2. Mechanical properties of diamond, TiB_2 and Si.

Materials	Young's modulus (GPa)	Poisson's ratio	Coef. of thermal expansion (K^{-1})
Diamond	1054	0.3	1.0×10^{-6}
TiB_2	550	0.2	6.4×10^{-6}
Si	150	0.3	4.2×10^{-6}

3.2.2 Analysis of Thermal Residual Stress

Finite element analysis (here, the triangle liner element method) was used to evaluate thermal residual stress near the interface of the Diamond/TiB_2/Si composites. Time- and temperature-dependent properties were ignored. The distributions of the radial stress in the 1.67 mm middle layer are shown in Figure 2.14 in MPa. In a no-middle layer system (n=0), was there a remarkable stress mismatch near the interface of two layers. On the other hand, in a seven-layer system (n=5), the relaxation of the stress mismatch near each interface was confirmed.

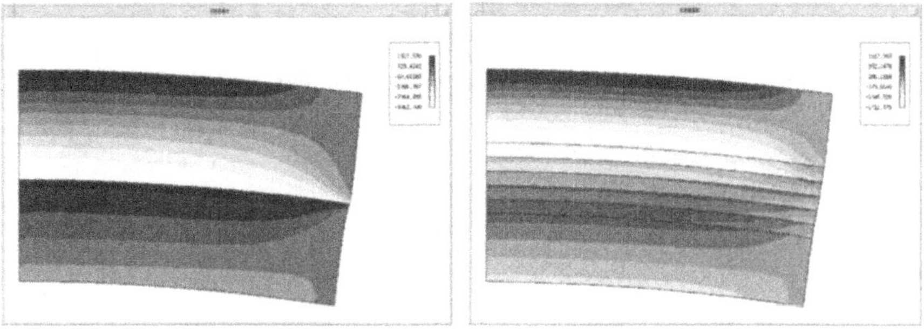

(a) n = 0 (b) n = 5

Figure 2.14. Distribution of radial stress for diamond gradually dispersed composites.

Figure 2.15 shows that equivalent stress was normalized by the stress of the no-middle layer system (n=0) when the number of laminas in the middle layer was changed. The maximum equivalent stress of three-layer system (n=1) was half of the two-layer system. In the 1.67 mm middle layer, when the middle layer was more than two laminas, their stresses were almost the same and became about 45% of the stress of the no-middle layer system. On the other hand, in the 3.33 mm middle layer, when the middle layer was more than four laminas, the maximum equivalent stress relaxed to about 20% as there was no middle layer. It was clear that the middle layer brings about an immediate result and a remarkable reduction of the residual stress mismatch with an increase in the number of laminas.

3.3 Microscopic Residual Strain

To evaluate microscopic residual stress, analysis by finite element method (two-dimensional plane strain analysis) and strain measurement by neutron diffraction were performed and compared. Figure 2.16 shows a SEM image of the WC/10wt% Ni fabricated using the SHS method. The gray parts in this figure are WC particles. Using the developed Image Mesh Generation System on this SEM image, a microscopic finite element model was generated. Table 2.3 shows the mechanical properties of the WC and Ni. Thermal residual stress for this microscopic model as calculated from the cool down from the manufacturing temperature (1773K) to room temperature (295K). The strain distribution of WC/Ni is shown in Figure 2.17.

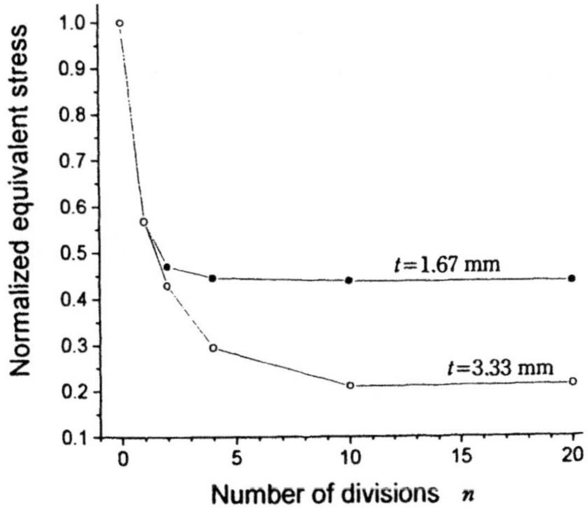

Figure 2.15. Relation between normalized equivalent stress and number of divisions n.

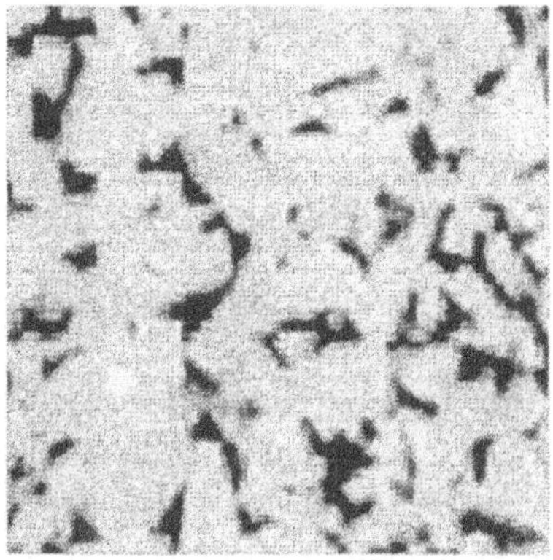

Figure 2.16. Digital SEM image of WC/10wt%Ni.

The actual inner residual strain for this material was measured by neutron diffraction. The experiment was performed using the Residual Stress Analyzer Equipment (RESA), a neutron diffraction apparatus, developed for strain measurement at the Japan Atomic Energy Research Institute (JAERI).

Table 2.3. Mechanical properties of WC and Ni.

Materials	Young's modulus (GPa)	Poisson's ratio	Coef. of thermal expansion (K^{-1})
WC	534. 4	0.22	6.5×10^{-6}
Ni	199. 5	0.312	1.3×10^{-5}

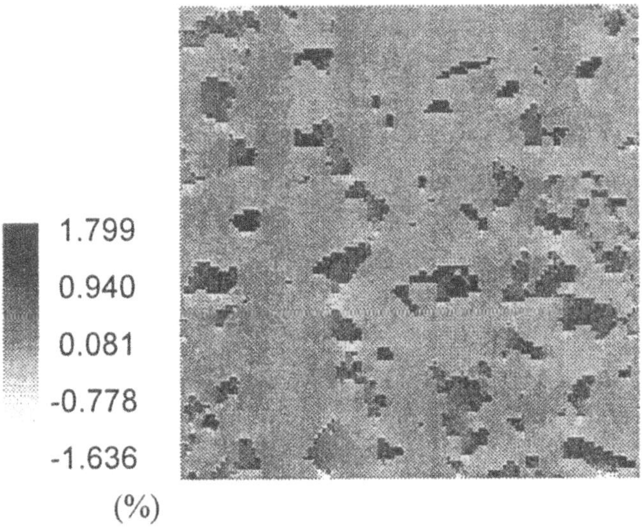

Figure 2.17. Distribution of microscopic strain (ε_x) of Ni part in WC/Ni.

In Figure 2.18, the black circles show the peak of the scattering angle of Ni(200), and the white circles show Ni(200) in WC/Ni. The neutron wavelength is 2.0887 Å. The continuous lines are Gauss-fitted curves. It has been proven that inter-atomic distance extends because the angle of diffraction of Ni in WC/Ni has deviated to the left. That is, it has been shown that tensile strain is being generated at Ni in WC/Ni. The strain of the Ni in the WC/Ni obtained from this measurement result was about 0.7%. The results of the residual strain obtained by microscopic analysis and the experiment are shown in Figure 2.19. The results of elastic microscopic analysis were in good agreement with the experimental results. Though the strain shows 0.7% and is a very high value, the particles in the WC/Ni layer cannot be plastic and appear to maintain an elastic condition.

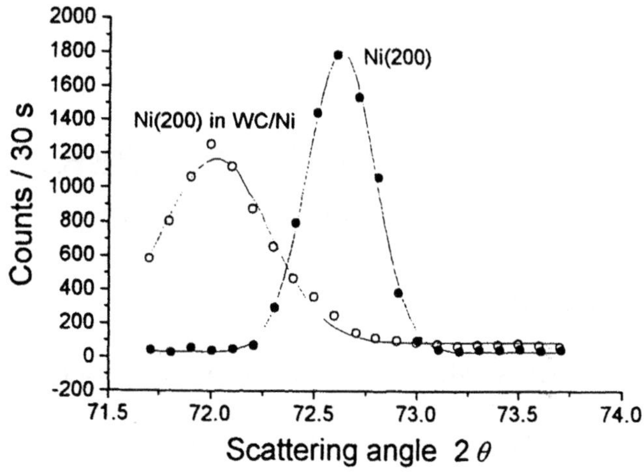

Figure 2.18. Peak of scattering angle of Ni(200).

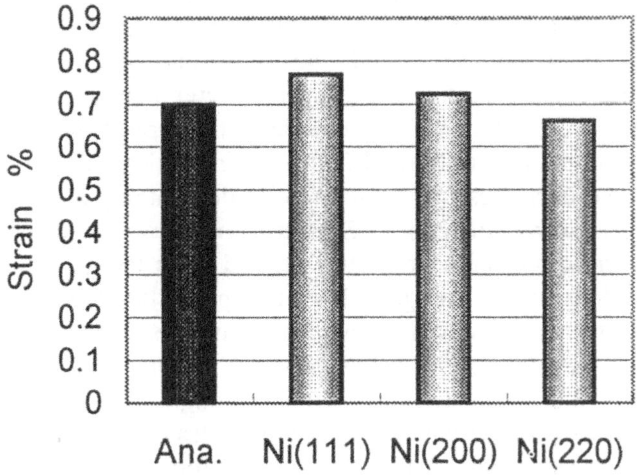

Figure 2.19. Comparison of analysis with experiment.

3.4 Conclusions

In this study, finite element analysis was used to evaluate thermal residual stress. In macroscopic analysis, the divisions of the middle layer tended to effectively reduce the residual stress mismatch near the interface of the laminas, but this was not particularly effective in more than ten layers. In microscopic analysis, the finite element model generated from the digital image was applied to calculate the microscopic strain. The obtained result was compared with the result measured using neutron diffraction. The results of the elastic microscopic analysis was in good agreement with the experimental results.

4. HIGH-TEMPERATURE PROPERTIES OF SURFACES AND INTERFACES

4.1 Introduction

Material properties reflect the microstructure, and as a result, processing conditions and methods must be chosen that yield "useful" microstructures. The concept of FGMs has expanded the range of useful microstructures to include: 1) multiphase materials in which the phase content and thus properties vary systematically with position, 2) systems in which chemical discontinuities are accompanied by continuous values of selected properties (e.g., thermal barrier coatings), and 3) single-phase materials, in which the composition does not vary with position, but processing conditions encourage the development of useful microstructural and property gradients. Hirai [18] has provided a review summarizing the various types of FGMs.

In conventional materials and FGMs, the energetic and kinetic properties of surfaces and interfaces play a vital role in determining the microstructure that develops during processing. Surface properties depend upon surface orientation; grain boundary properties depend upon the grain-to-grain misorientation. Both surface and grain boundary properties can also be affected by material chemistry, and thus, impurities can drastically alter the path of microstructural evolution. Impurities can alter the energetics of surfaces and interfaces, and thereby the thermodynamic driving forces for microstructural change. Dopants can also change the relative rates of grain boundary, lattice and surface diffusion, and thereby the densification-coarsening competition in powder-based processing methods [19]. Situations in which a difference in surface orientation, stability, and structure completely changes the mechanism and kinetics of morphological change can also arise [20]. Separating such chemical, energetic, and mechanistic effects in experiments utilizing powders is, at best, difficult, and more likely, impossible. This paper provides an overview of methods that allow fundamental studies of surface and interface properties in ceramics, and other materials as well.

4.2 Experimental Methods

The use of lithography to produce controlled-geometry surface cavities in substrates of known composition and/or crystallographic orientation, and their ultimate transformation into internal intragranular or intergranular cavities by hot pressing has been discussed in detail in prior work [21, 22]. This general method can also be used to generate defects at single-crystal/polycrystal interfaces, and at interfaces between dissimilar materials, e.g., ceramic/metal interfaces. Early work utilizing this approach focussed on sapphire and alumina. More recent work, notably that of Narushima [23], has extended the method to silicon carbide ceramics.

Such microdesigned internal defect structures provide an ideal starting point for studies of morphological evolution and shape equilibration at elevated temperature [e.g., 22]. Lithography enables duplication of the defect geometry in substrates of different crystallographic orientation, allowing crystallographic effects to be isolated. Identical defect arrays can also be introduced in materials in which the composition has been modified by dopants; when the surface orientation is maintained, this allows systematic study of impurity effects on the energetic or kinetic properties of surfaces. When optically transparent materials are used, optical microscopy provides an attractive means of monitoring morphological changes nondestructively. Such low resolution methods of interrogation can be coupled with higher spatial resolution techniques such as sem, afm, and tem applied to re-exposed surfaces, cross-sections, and foils containing such internal defects.

Although fundamental studies of surfaces are vital, fundamental studies of solid-solid interfaces, grain boundaries, are equally critical. There are basically two approaches to studying grain boundaries. One approach focuses on polycrystalline material, measures the properties of many boundaries, and often assesses the properties of the "average" grain boundary. The alternative approach is to perform fundamental studies on bicrystals. A major difficulty with this latter approach has been the limited availability, high-cost, or difficulty of fabricating suitable bicrystals. Recently, Kitayama *et al.* have produced sapphire bicrystals [24] by extended a method originally formulated by Scott [25], and later adapted to the fabrication of glass/ceramic interfaces by workers at NIST [26]. This method relies on tandem epitaxial growth of single crystal seeds into a thin sacrificial polycrystalline layer prepared by conventional means. In work by Mammana [24], suitably misoriented sapphire seeds consumed 100 to 500 μm thick 500 ppm, titanium-doped polycrystalline alumina substrates during anneals at 1600°C to produce 0-2°, 20° and 40° basal twist boundaries. Bicrystals prepared by this method have just recently been examined using atomic resolution tem, and initial indications are that boundaries free of any obvious (contaminant) second-phases have been produced. Further details of sample preparation and tem characterization will be published elsewhere [27]. Current efforts are focused on extending this approach to silicon and silicon carbide, as well as to model metallic systems.

The combination of these two experimental methods provides an integrated suite of tools for assessing the properties of surfaces and interfaces. The following provides a short account of some of the recent findings of the lithography-based work, and focus of the on-going bicrystal research.

4.3 Surface Energetics in Undoped and Doped Sapphire

Kitayama has used the lithographic method described to generate arrays of small intragranular cavities in both undoped and Ca-, Mg-, and Ti-doped sapphire. Sapphire substrates with such internalized defects were annealed for prolonged periods of time at 1600°C and 1800°C until near-equilibrium (Wulff) shapes were achieved. Careful sectioning and polishing of the samples was successful in re-exposing these cavities, allowing detailed measurements of the pore shape. Since the orientation of the single crystal sapphire is known, AFM interrogation of the pore surfaces provides information on the crystallographic orientation of the facets. Five stable planes (**c**, **s**, **r**, **a**, and **p**) were identified in undoped alumina, and the surface energies relative to the basal c (0001) surface were determined, as summarized in Table 2.4. Some dopants (Mg^{2+}, Ti^{3+}) modify the relative energies of these five stable planes, while others (Ca^{2+}, Ti^{4+}) cause the appearance or disappearance of planes [28].

Table 2.4. Experimentally determined relative surface energies of undoped α-alumina.

Plane	1600°C Choi *et al.* [28]	1600°C Kitayama [27]	1800°C Kitayama [27]
c(0001)	1.0	1.0	1.0
s$(10\bar{1}1)$	1.07 ± 0.02	0.947 ± 0.016	1.042 ± 0.019
r$(10\bar{1}2)$	1.05 ± 0.02	0.855 ± 0.017	0.950 ± 0.030
a$(11\bar{2}0)$	1.22 ± 0.05	0.947 ± 0.026	1.080 ± 0.017
p$(11\bar{2}3)$	1.06 ± 0.02	0.957 ± 0.026	1.029 ± 0.016
m$(10\bar{1}0)$	>1.16	>1.008	>1.115

4.4 Kinetic of Morphological Change in Undoped Sapphire

The rate at which fully facetted nonequilibrium shaped particles and pores approach their equilibrium (Wulff) shape via surface diffusion was modeled, and calculations relevant to alumina were performed[29]. Models appropriate to 2-D features, that consider initial particle/pore shape, size, surface energy anisotropy, and temperature (surface diffusivity) as variables [30], and models appropriate to 3-D features [31] were developed. In each case, two approaches to modeling the surface flux were explored, and found to yield nearly identical results. The most important parameters affecting the evolution times are the cross-sectional area, and the temperature through its effect on the surface diffusivity. Importantly, pores of micron size are predicted to reach near-equilibrium shapes in reasonable times at temperatures as low as 1600°C.

To assess the applicability of the models, nonequilibrium shaped pores of identical initial geometry (≈ 20μm × 20μm × 0.5μm) were introduced into (0001), $\{10\bar{1}2\}$, $\{11\bar{2}0\}$, and $\{10\bar{1}0\}$ surfaces of sapphire single crystals using micro-fabrication techniques, ion-beam etching and hot-pressing. The large (≈20μm × 20μm) faces of the pore were low-index surfaces whose nature is dictated by the wafer orientation. Pore shape evolution during anneals at 1900°C was monitored. The kinetics of shape evolution are highly sensitive to the crystallographic orientation and stability of the low-index surface that dominates the initial pore shape. The measured variations of the pore shape were compared to the model predictions. The observations suggest that when the initial bounding surface is unstable, shape relaxation may be controlled by diffusion. However, surface attachment limited kinetics (SALK) appear to play a major role in determining the pore shape evolution rate in cases where the initial bounding surfaces have orientations that are part of the Wulff shape.

4.5 Pore Shape Equilibration in 6H Silicon Carbide

Narushima [23] has adapted these experimental approaches to investigate the morphological evolution of nonequilibrium shape pores introduced into the (0001) surface of 6H silicon carbide. The morphological evolution of these internal cavities (negative crystals) in response to anneals of up to 128 h duration at 1900°C was examined using optical microscopy. Surface energy anisotropy and facetting have a strong influence on both the geometric and kinetic characteristics of evolution. Decomposition of $\{1\bar{2}10\}$ cavity edges into $\{10\bar{1}x\}$ facets are observed after 16 h anneals, indicating that $\{1\bar{2}01\}$ facets are observed after 16 h anneals, indicating that {1201} faces are not components of the Wulff shape. The shape evolution kinetics of penny-shaped cavities were also investigated. Experimentally observed evolution rates decreased much more rapidly than predicted by a model in which surface diffusion is assumed to be rate-limiting. This suggests that the development of facets, and the associated loss of ledges and terraces during the initial stages of evolution results in an evolution process controlled by surface attachment limited kinetics (SALK). These results thus reinforce the coupling between surface energetics and kinetics seen earlier in sapphire.

4.6 Bicrystals via Epitaxially Stimulated Transformations (BEST)

Prior work by Powers and Glaeser [32] demonstrated that Ti additions to alumina can lead to increased rates of grain boundary migration. In view of the long-standing debate regarding the possible role of liquid phases in both the sintering and grain growth of Ti-doped sapphire, dense (>99%) polycrystals were prepared by doping a high-purity (99.997%) alumina powder. Using the method described previously, a series of bicrystals was produced. The characterization of these specimens is at an early stage. Nonetheless, the results are sufficient to show that no obvious second phase exists along the grain boundary, suggesting a solid-solution effect of titanium. Chemical characterization shows evidence of titanium segregation. A complete analysis of the misorientation, grain boundary structure, and grain boundary chemistry is in progress. The most important finding of this and closely related studies on silicon and copper alloys is that bicrystals can be fabricated simply and quickly using this method. This suggests that the approach may have broader utility and provide a useful tool for studying grain boundary misorientation-structure-chemistry-property relationships.

REFERENCES

1. Watanabe, R., Kawasaki, A. and Murahashi, N., Asso, J. Mater. Eng. Resour., (in Japanese), 1 (1988), 36.
2. Watanabe, R., Micromeritics, (in Japanese), 33(1989), 76.
3. Watanabe, R., MRS Bulletin, 20-1(1995), 32.
4. Suresh, S. and Mortensen, A., *Fundamentals of Functionally Graded Materials*, Institute of Materials, UK, (1998).
5. Miyamoto, Y., Kaysser, W.A., Rabin, B. H., Kawasaki, A. and Ford, R. G., *Functionally Graded Materials: Design, Processing and Applications*, Kluwer Academic Publishers, (1999).
6. Shinagawa, K., in *Functionally Graded Materials 1996*, edited by Shiota, I. and Miyamoto, Y., Elsevier, (1997), 69.
7. Riedel, H. and Kraft, T., Materials Science Forum, 308-311(1999), 1035.
8. Shinagawa, K. and Hirashima, Y., Materials Science Forum, 308-311(1999), 1041.
9. Shinagawa, K. and Hirashima, Y., Proc. Plasticity '2000, (2000), (in press).
10. Awaji, H., Takenaka, H.,Honda, S. and Nishikawa, T., J. Ceram. Soc. Japan, 107(1999), 780-785.
11. Awaji, H., Takenaka, H., Honda, S. and Nishikawa, T., Trans. Jpn. Soc. Mech. Eng.,

65A(1999), 2318-2324 (in Japanese).

12. Sivakumar, R., Takenaka, H., Honda, S., Nishikawa, T. and Awaji, H., Proc. 16th International Japan-Korea Seminar on Ceramics, Nov. 30- Dec. 2 (1999) Okayama, 324-28.
13. Tanigawa, Y. ,Oka, N., Akai, T. and Kawamura, R., JSME Int J., 40 (1997), 117-127.
14. Tanigawa, Y., Appl. Mech. Rev., 48 (1995), 287-300.
15. Carslaw, H. S. and Jaeger, J. C., *Conduction of Heat in Solids*, 2nd ed., Oxford Univ. Press, (1959).
16. Ohyanagi, M., Tsujikami, T., Sugahara, S., Koizumi, M., Levashov, E. A. and Borovinskaya, I. P., Graded Materials of Dispersed Diamond TiB_2-Si Composite Fabricated by SHS/Dynamic Pseudo Isostatic Compaction, Proceeding of the 5th International Symposium on Functionally Graded Materials, October 26-29 (1998), Dresden, Germany, Materials Science Forum, Trans Tech Publications, 308-311(1998), 145-150.
17. Tsujikami, T. and Suzuki, K., Development of Image Mesh Generation System for Composite Materials, The First Asian-Australasian Conference on Composite Materials, October 7-9 (1998), Osaka, Japan, 310.
18. Hirai, T., Chapter 20 in Materials Science and Technology: A Comprehensive Treatment, R. J. Brook, Ed. Weinheim, Germany: VCH Verlagsgesellschaft mbH, (1996).
19. Powers, J. D., Ph. D. thesis, msme, University of California, Berkeley, (1997).
20. Kitayama, M. and Glaeser, A.M., Mater, J. Synth. Proc., 6-3, (1998), 161-167.
21. Rödel, J. and Glaeser, A. M., Yogyo Kyokai Shi, 99-4 (1991), 251-265.
22. Kitayama, M., Powers, J. D., Kulinsky, L. and Glaeser, A M., J. Eur. Ceram. Soc., 19-13-14 (1999), 2191-2209.
23. Narushima, T. and Glaeser, A. M., unpublished research, (1999).
24. Mammana, E., Radetic, T. ,Gronsky, R.and Glaeser, A. M., unpublished research, (1999).
25. Scott, W. D., Trans. Brit. Ceram. Soc., 66-7 (1967), 315-318.
26. Kim, D. Y., Wiederhorn, S. M., Hockey, B., Handwerker, C. A. and Blendell, J. E., J. Am. Ceram. Soc., 77-2 (1994), 444-453.
27. Mammana, E., Taylor, S. Radetic, T. Gronsky, R. and Glaeser, A. M., unpublished research, (2000).
28. Kitayama, M., Ph.D. thesis, msme, University of California, Berkeley, (1996).
29. Choi, J., Kim, D., Hockey, B. J., Wiederhorn, S. M., Handwerker, C. A., Blendell, J. E., Carter, W. C. and Roosen, A. R., J. Am. Ceram. Soc., 80-1 (1997), 62-68.
30. Kitayama, M., Narushima, T., Carter, W. C., Cannon, R. M. and Glaeser, A M. in press, J. Am. Ceram. Soc., (2000).
31. Kitayama, M., Narushima, T. and Glaeser, A M., in review, J.Am. Ceram. Soc., (2000).
32. Powers, J. D. and Glaeser, A. M., Ceram. Eng. Sci. Proc., 18-4 (1997), 617-623.

Chapter 3

PROCESSING AND SYSTEM

1. NEW POWDER METALLURGY

1.1 One-Step Synthesis and Densification by Electric Field Activation

1.1.1 Introduction

Formation of functionally-graded materials (FGMs) by self-propagating high-temperature synthesis (SHS) has been amply demonstrated in many previous studies [1]. Layers of reactant powders are stacked in a compositionally-graded sequence and ignited at one end to initiate a self-sustaining combustion wave to fabricate the FGM. However, FGM formation by this method is subject to the same thermodynamic and kinetic limitations of the SHS process itself [2], and as such the synthesis of a relatively large number of material systems (monolithic, composites, and FGMs) is excluded. The use of an electric field to overcome these limitations, and hence to activate the synthesis process, has been demonstrated in previous investigations for many materials, including FGM systems [3-8].

More recently, the process of field activation has been modified to make possible the simultaneous synthesis and densification of materials, including FGMs [9, 10]. Typically elemental reactants are subjected simultaneously to a high-density current and a uniaxial pressure. The generally accepted role of the current in initiating the reaction is shown schematically in Figure 3.1. The current raises the temperature of the reactants to the ignition temperature, T_I, at a rate which is assumed to be dictated solely by Joule heating. But once the reaction commences, the rate of temperature increase is now due to both electrical and chemical energies. While in principle other techniques (e.g., reactive hot-pressing) can be used to achieve the same goal, the crucial difference between the present method and the others is in the rate of heating. Lower heating rates (as achieved in furnace heating) often result in the formation of pre-combustion phases (through interfacial diffusion) which can alter the mechanism of the process [11]. In the initial work on field activation AC currents were utilized in a Field-Activated Pressure-Assisted Synthesis apparatus (FAPASA). But recently the work has been carried out in a Spark Plasma Sintering (SPS) apparatus [12].

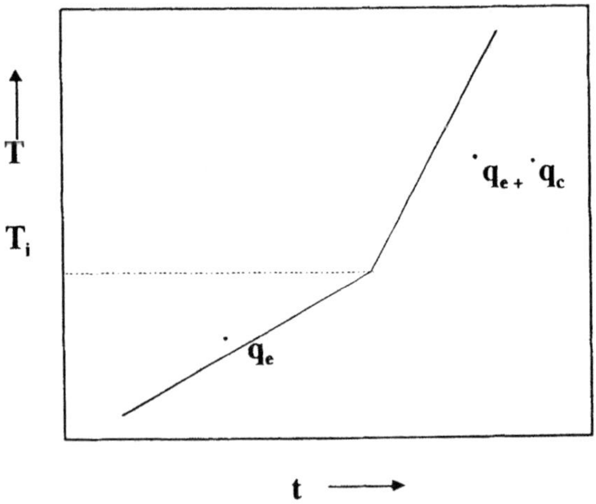

Figure 3.1. Schematic representation of thermoelectric ignition.

In this section Munir *et al.* present results of investigations on the simultaneous synthesis and densification of composite and FGM materials using the process of field activation. In order to provide an understanding for the role of the current in this process, modeling studies were also carried out and are briefly discussed at the outset.

1.1.2 Modeling of Current Induced Synthesis Reactions

As implied in Figure 3.1, the passage of a current through reactants raises the temperature to a point where the entire sample reacts, a process referred to variously as simultaneous or volume combustion. However, observations have been reported indicating that this may not be the case and that a reaction in a waveform may take place under these circumstances [13,14]. Aside from a fundamental interest in this phenomenon, there is also a practical significance. Waveform reactions are associated with steep temperature gradients and can therefore result in compositions and structures which are non-equilibrium in nature. Since it is not possible to observe such waves in the FAPASA and SPS experimental systems, Munir *et al.* investigated the process through two-dimensional simulation [15].

Results from the two simulated systems (SiC and $MoSi_2$) show that both volume and waveform combustion take place during synthesis. When the cylindrical samples are small, the combustion process proceeds in a volume mode and when the samples are large, the combustion proceeds in a waveform. Figures 3.2 shows conversion, η, profiles during the synthesis of SiC for samples with radii of 1.0 and 2.0 cm, respectively [15]. For the smaller samples, Figure 3.2(a), the conversion profiles are relatively flat, indicating a broad reaction zone and relatively flat temperature gradients. In contrast, for the larger samples, Figure 3.2(b), the profiles are characteristic of a wave combustion process with the attendant steep temperature profiles. A more revealing result of the modeling work relates to the distribution of the current during synthesis. Figure 3.3 shows current density profiles for the 1 and 2-cm samples in the case of SiC synthesis. In the former, the current is initially distributed broadly through the sample but then becomes somewhat localized during the end of the combustion process when the sample has nearly totally converted to SiC. Since the product (SiC) is not as good an electrical conductor as the reactants, it carries less current and thus at an advanced conversion level the current becomes progressively more confined, as seen in the figure. In sharp contrast to these are the current profiles for the larger samples, Figure 3.3. Here, initially the profiles are broad but quickly become highly localized and move along the sample in concert with the combustion wave. These latter observations are consistent with results of modeling on SHS waves in SiC, indicating that in this case the current is primarily confined to the combustion zone [3]. The practical significance of this is that the electrical energy needed to activate the process is primarily utilized in the reaction front.

Similar analysis was made of the synthesis of $MoSi_2$ under conditions simulating the SPS process. The results are qualitatively similar to those of SiC, except that the transition from volume to waveform combustion takes place at a larger sample size, in this case as the radius of the sample is increased from 2 to 3 cm. The role of sample size in the transition from volume to waveform combustion mode is a consequence of the thermo-physical characteristics of the reactants and product and the boundary conditions imposed by the heat loss at the sample surface. It should be pointed out that the modeling work described above does not take into

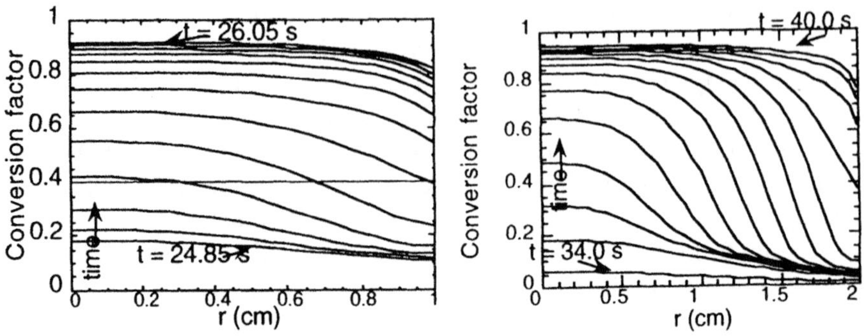

(a) (b)

Figure 3.2. Simulation results of conversion profiles during SiC synthesis: (a) 1 cm radius sample and (b) 2 cm radius sample.

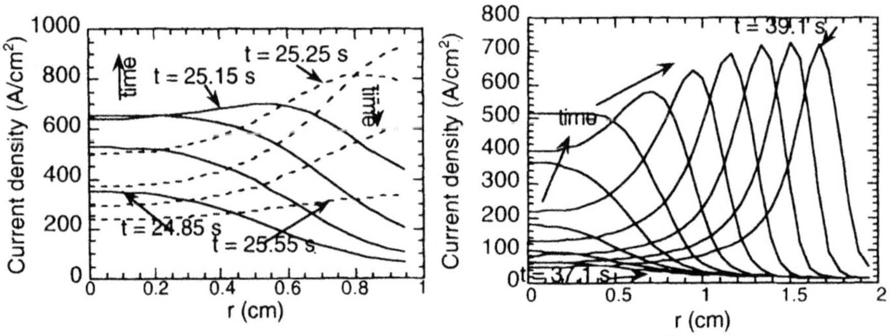

(a) (b)

Figure 3.3. Simulation results of current density profiles during SiC synthesis: (a) 1 cm radius sample and (b) 2 cm radius sample.

account the presence of a die, the typical experimental configuration in SPS and FAPASA experiments. The common use of graphite dies represents an electrical analogy of two parallel circuits, one of the sample and the other of the die. The effect of the current therefore depends on the relative conductivities of the sample and the die and their dependence on temperature. The incorporation of a die in these simulations is being investigated currently to evaluate the role of the conductivity of the die in the SPS process. Both conducting and non-conducting dies will be considered in the simulation.

In the modeling study described above, the role of the field (or current) has been implicitly assumed to be thermal, i.e., adding Joule heat to the reaction. While Joule heating may be a significant element of the role of the field, it is likely not to be the only contribution. Mass transport processes can be affected by the current through electromigration which arises from the "electron wind" effect [16]. In the presence of high density DC currents the atomic flux across an interface (e.g., between contacting particles in the SPS) is enhanced by this effect [17] and thus can contribute to the synthesis process in a unique way.

1.1.3 Simultaneous Synthesis and Densification of FGMs

As indicated above, the application of a uniaxial pressure during field activated combustion has been shown to produce high density products from, typically, elemental reactants [18]. The method has been recently used to investigate the formation of FGMs in the system $MoSi_2$-SiC [10]. Elemental powders (Mo, Si and C) were mixed in proportions to produce $xMoSi_2$-(1-x) SiC FGMs with $0 < x < 1.0$ and in layers with increments (Δx) of 0.1. To better understand the formation of the FGM, the initial experiments focused on the formation of individual layers (i.e., composites) and bilayers of $xMoSi_2$-(1-x)SiC. The mixed powders were placed in a graphite die (one layer on top of another for bilayers) and reacted in the FAPASA or SPS facilities. In the FAPASA the samples were subjected to a high AC current (560 $A.cm^{-2}$) and a pressure of 95 MPa. In the SPS apparatus, a pulsed DC current used was at a density of 1034 $A.cm^{-2}$ and the uniaxial pressure was 63 MPa. Under these conditions, the temperature increased to a maximum of 1350 to 1450°C (depending on composition) in about 6 min in the FAPASA and to 1450°C in 10 min in the SPS. The samples were held at these temperatures for 5 min and then allowed to cool naturally. Indication of a reaction was signaled by a sudden volumetric displacement, probably initiated by liquid phase formation.

X-ray diffraction analysis showed that the product contained the desired phases ($MoSi_2$ and SiC) but in some cases it also contained the phase $Mo_{4.8}Si_3C_{0.6}$. The relative amount of this phase depended on the x value, i.e., on the amount of carbon in the reactants. For x= 3, the products contained significant amounts of $Mo_{4.8}Si_3C_{0.6}$, but samples with $x \geq 0.5$ showed no XRD evidence of the presence of this carbosilicide.

The demonstrated success of formation of dense composites in one step was followed by investigation on the formation of $xMoSi_2$- (1-x)SiC FGMs. Layered materials ranging from pure $MoSi_2$ to pure SiC in x = 0.1 increments could be synthesized by this method under conditions similar to those utilized for the individual layers. Bilayers and multilayers were successfully made. Figure 3.4 shows the concentration profile across an interface of a bilayer with x = 0.9 and x = 0.5. The results show that the interface has an approximate thickness of 100μm. Multilayer FGMs spanning the entire range could be made in one step. The resulting FGM, however, could not be made with uniform density. SiC-rich layers were found to contain significant porosity. Density measurements showed an abrupt decrease in relative density occurs as x decreases below 0.5. Since Munir *et al.* intended to characterize the FGMs mechanical properties, the subsequent work focused only on high density FGMs, i.e., those with $0.5 < x < 1.0$ FGMs with these compositional limits had densities in the range 94 to 97%. A typical FGM is shown in Figure 3.5 which is an optical micrograph of a layered sample with x ranging from 0.5 to 0.9. The figure shows a porous region between the 60 and 70% Si layers; the reason for the formation of this porosity and its localization to the interface between these specific layers is not understood. Figure 3.6 depicts results of electron microprobe analysis and SEM images of an FGM with a compositional range x=1.0 to x=0.60. The composition is plotted as wt % S across the five layer $MoSi_2$-SiC FGM.

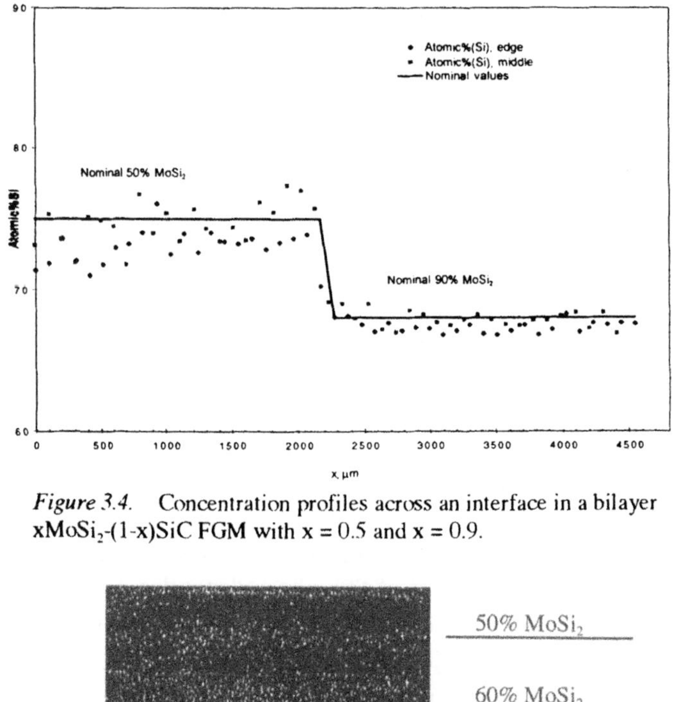

Figure 3.4. Concentration profiles across an interface in a bilayer $xMoSi_2$-$(1-x)SiC$ FGM with x = 0.5 and x = 0.9.

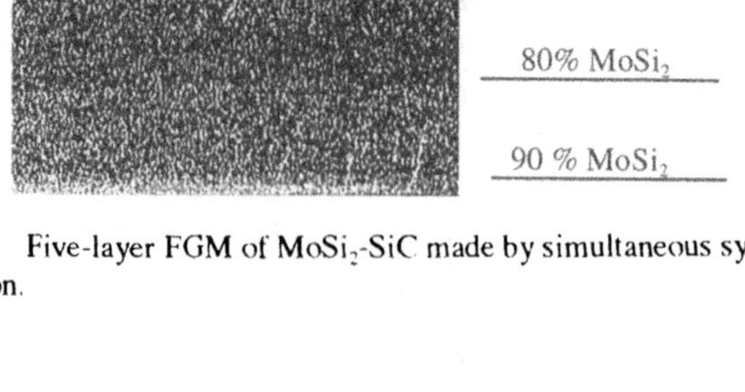

Figure 3.5. Five-layer FGM of $MoSi_2$-SiC made by simultaneous synthesis and densification.

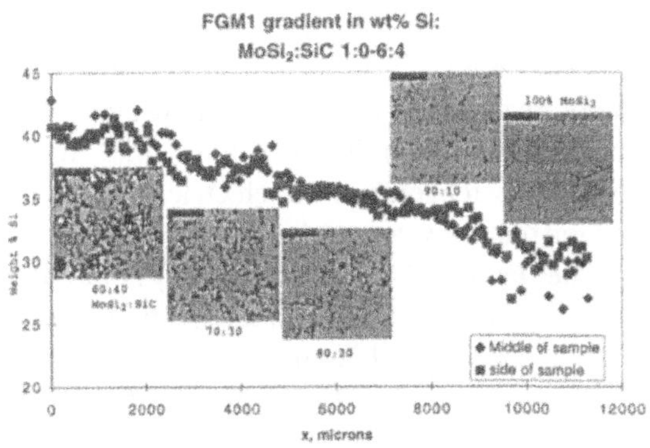

Figure 3.6. Concentration profiles and SEM images in a five-layer $MoSi_2$-SiC FGM.

1.1.4 Concluding Remarks

The feasibility of a one-step synthesis and consolidation of a ceramic-ceramic FGM was investigated. FGMs of $xMoSi_2$-(1-x)SiC were synthesized from elemental reactants using a field-activated pressure assisted combustion synthesis method and utilizing FAPASA and SPS facilities. It was demonstrated that FGMs with $0.0 \le x \le 1.0$ could be made by this approach, but that layers with $x < 0.5$ contained significant porosity. For FGMs with $0.5 \le x \le 1.0$, the density was in the range 94 to 97%. XRD results showed the formation of only the two expected phases for samples with $x \ge 0.5$. For samples with lower x values, the product also contains $Mo_{4.8}Si_3C_{0.6}$ in significant amounts. Microstructural and compositional profile analyses were made on the resulting FGMs.

To understand the role of the field (current) on synthesis, modeling studies were made. A brief summary of the simulation results is presented.

1.2 Fabrication Using Spark Plasma Sintering

1.2.1 Introduction

Consolidation by spark plasma sintering (SPS) is different from that by pressureless sintering and hot pressing [19-21]. The electric source of SPS is similar to that of an electric discharge machine that generates a direct pulse current. This current creates unique effects on materials: (1) spark plasma, (2) electric field, (3) electric current in conductor or skin current on semiconductor and insulator, (4) impact of spark plasma, (5) direct heating. Spark plasma (1) is considered to be effective on the reaction and etching of organic materials and the elimination of adsorbed gases from powder, and the rupture of films coated on the powder surface. Advanced materials can be synthesized on the basis of spark plasma. Functionally graded materials (FGMs), which cannot be obtained by other methods, can be fabricated using this effect. The electric field (2) is associated with the acceleration of diffusion of materials and vacancies in the electric field. The electric current in conductor or skin current on semiconductor and insulator (3) is significant for crystal growth. The impact of spark plasma (4) is due to the mechanical pressure caused by the spark plasma. Direct heating (5) is related to the fabrication process under graded temperatures. Dense FGMs can be fabricated from materials with different sintering or consolidating temperatures.

Polyimides are heat-resistant polymers and are used over 400°C. Two kinds of polyimide are synthesized: the one is thermoplastic, and the other is thermosetting. Thermosetting polyimide is not consolidated by heat. The spark plasma can excite a couple of chemical bonds of the polyimide, and make new chemical bonds to transform it into a dense bulk [22]. It has been impossible to prepare a polymer/metal FGM because the consolidation temperature is different for these two materials. Thermosetting polyimide is consolidated near 400°C, and Cu is sintered at 520°C by SPS. SPS makes it possible to fabricate a dense polyimide/Cu FGM.

Tungsten carbide has not been sintered without sintering aids by heating. The hard metal of WC is prepared by adding sintering aids of Co, Ni and Fe [23]. Pure WC powder can be sintered by SPS [24]. Pure WC compacts have been characterized by higher hardness and lower toughness [25]. The WC/Mo and WC/WC (Co) FGMs from pure WC were fabricated by SPS.

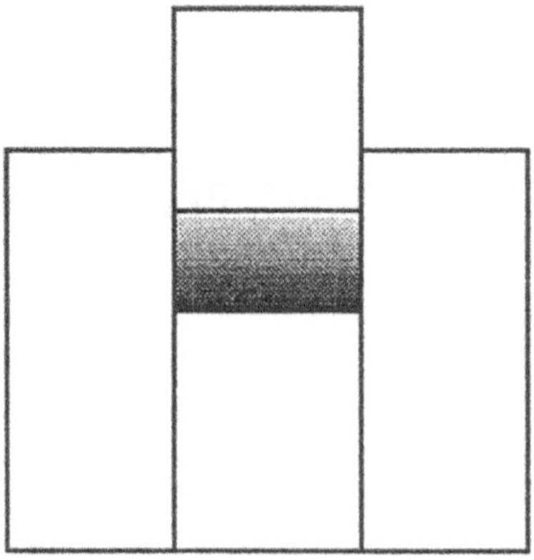

Figure 3.7. Die arrangement for graded temperature.

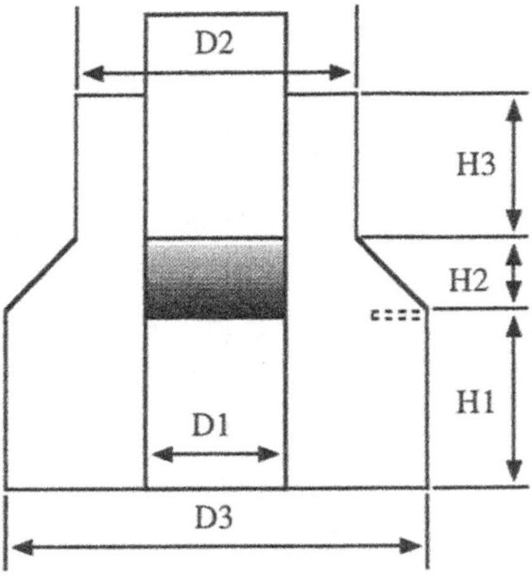

Size of die

D1 (mm)	D2 (mm)	D3 (mm)	H1 (mm)	H2 (mm)	H3 (mm)
20	40	55	20	15	20
20	40	60	25	10	20
50.4	80	110	50	20	20
80.4	110	150	50	20	20
100.4	130	180	50	20	20

Figure 3.8. Die shape with different diameter.

1.2.2 Preparation and Evaluation

The materials examined were Cu (Nippon Atomized Powders Co. Ltd., 5μm), thermosetting polyimide (Kaneka Co. Ltd.), WC (Japan New Metals Co. Ltd., 0.4μm), Mo (Tokyo Tungsten Co. Ltd., 3.8μm), and Co (Kokkola Chemicals, 1.5μm) powders. The equipment employed for the consolidation of the thermosetting polyimide and fabrication of the FGM was the spark plasma sintering system (Izumi Technology Co. Ltd., SPS-1050). The consolidation of the thermosetting polyimide at homogenous temperatures was carried out from 9.8 to 147 MPa using graphite and hard alloy straight dies as shown in Figure 3.7. The thermosetting polyimide and Cu powders were mixed. Polyimide, Cu and mixed powders were layered in the die shown in Figure 3.8 (D1=20, D2=40, D3=70, H1=20, H2=15, H3=20 mm). Temperature and pressure schedules for the polyimide/Cu FGM formation are illustrated in Figure 3.9.

WC and Mo powders were mixed using an agate mortar and a pestle for 0.5 h. WC, Mo and the mixed powders were layered in a graphite die shown in Figure 3.7, and sintered at 39.2 MPa in a vacuum. The sintering temperature was raised from room temperature to 1500°C over 20 min. then the temperature of 1500°C was held for 5 min.

WC plates were sintered at 1800°C by SPS. The plate was put in a straight graphite die, and then the mixed powder of WC and Co at 10 wt% was added to it. The straight die was heated from room temperature to 1250°C over 15 min and then held at 1250°C for 5 min in a vacuum to obtain a WC/WC (Co) FGM. The sintering pressure was 20 MPa.

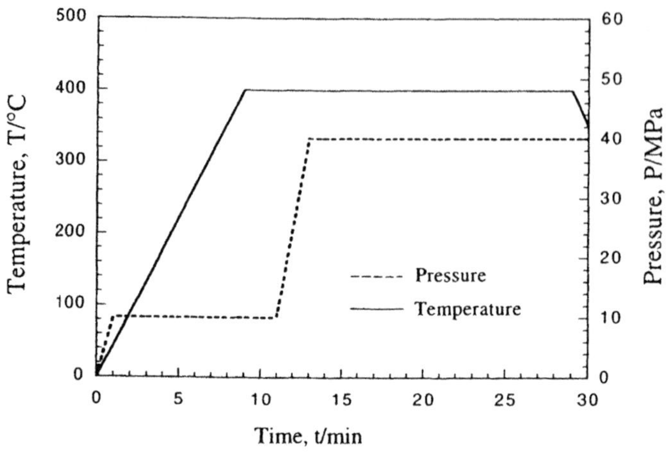

Figure 3.9. Temperature and pressure schedules for FGM formation.

The consolidated samples were polished, and their properties were measured. The polished surfaces were observed with an optical microscope (Olympus, SZH). The microstructure was analyzed with a scanning electron microscope (JEOL, JXA-8621) with microprobe analysis (EPMA). The elastic constants were determined by a pulse-echo overlap ultrasonic technique using an ultrasonic detector (Hitachi Kenki, ATS-100) and a storage oscilloscope (Iwasaki Tsushinki, DS6411). A Vickers hardness tester (Akashi, MVK-E) was used for the measurement of hardness and toughness of the composites and FGMs under a load of 98 N. Fracture toughness was computed from Vickers hardness, Young's modulus and crack lengths at the indentations [26].

1.2.3 Graphite Die Creating Graded Temperature

The graphite dies of SPS play two roles: the one is the sintering mold, and the other is the heating element. The heating element of the graphite dies is suitable for rapid heating and cooling, because their heat capacity is low without insulation and a large heating element. Graded temperature can be obtained according to the difference of electric resistance due to the asymmetrical shape and alignment of the graphite dies as shown in Figure 3.7. The arrangement of a die and punches shown in Figure 3.7 creates the graded temperature, using the high temperature of the upper punch [27]. The die with different diameters shown in Figure 3.8 results in graded temperature for sintering, and the temperature gradient is changed by varying the sizes of the dies [28-31].

The maximum temperature difference between the upper and lower parts of the powder sample in the die is 100 to 200°C below 1000°C, and 100 to 300°C above 1000°C. The difference changes according to the sintering processes, because it depends on the density and thermal conductivity of the powder. FGMs can be fabricated from materials with different sintering temperatures using these dies.

1.2.4 Polyimide/Cu FGM

Practical metals such as Al, Cu, Ni, Fe etc. are sintered at more than 400°C. Most polymers are decomposed before they reach the sintering temperature. The sintering temperatures of Ni and Fe are too high to be combined with polymers. Heat-resisting polymers make it possible to prepare composites with Al and Cu. Thermoplastic polyimide softens near 325°C, and is used for a polyimide/Al FGM [29]. Thermosetting polyimides can be used over 400°C, and for polyimide/Al FGMs [30].

1.2.4.1 Consolidation of Thermosetting Polyimide

The spark plasma worked at a lower temperature before completing consolidation. The consolidation temperature of the thermosetting polyimide depended on the pressure, as shown in Table 3.1. The polyimide, which was consolidated at 380°C and at 19.6, 29.4, and 39.2 MPa, was partially carbonized and became dark and brittle. Polyimides are heat-resistant over 400°C, and cannot be carbonized at this temperature. The energy of the plasma was larger at high pressure than at low pressure. To avoid carbonization, the polyimide was at first consolidated at a lower pressure such as 10 MPa. The consolidated polyimide was not carbonized at the high temperature of 400°C and at the high pressure of 40 MPa [22].

Table 3.1. Consolidation of polyimide.

	Pressure (MPa)				
	9.8	19.6	29.4	39.2	147
Consolidation temp. (°C)	380	290	290	215	200
Bulk density (g/cm³)	1.06	1.25	1.33	1.34	1.44
Poisson's ratio	0.305	0.345	0.360	0.359	0.397
Young's modulus (GPa)	1.45	3.00	3.15	3.17	4.40

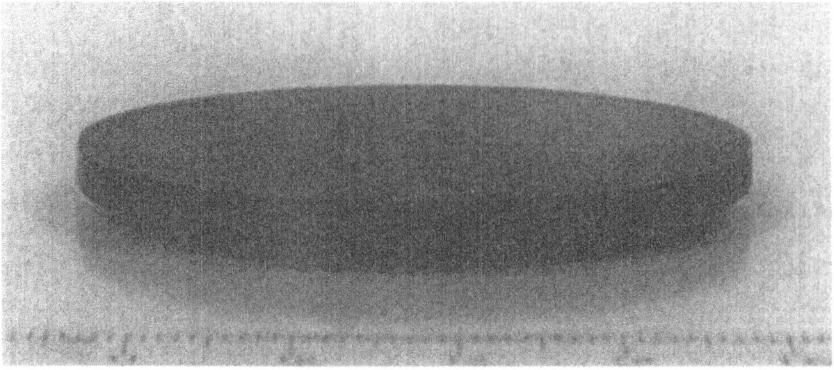

Figure 3.10. Polyimide/Cu FGM disk of 50 mm in diameter.

1.2.4.2 Polyimide/Cu FGM

The temperature and pressure schedules for the FGM formation are illustrated in Figure 3.9. The thermosetting polyimide was consolidated at first at 10 MPa and at 400°C, and then the Cu powder was sintered at the same temperature by increasing the pressure up to 40 MPa. The two layers inserted between the polyimide and Cu layers were not enough to prevent cracking of the FGM at the layer interfaces. A polyimide/Cu FGM free from cracks and pores was fabricated by inserting three layers of 75/25, 50/50 and 25/75 of polyimide/Cu vol %. The effective temperature was obtained and controlled near the polyimide layer. The Cu powder was sintered at 520°C and at 39.2 MPa, and the polyimide was consolidated at 400°C and at 9.8 MPa. The temperature difference between the polyimide and Cu layers was about 120°C. The thermal expansion coefficient of the Cu is 17×10^{-6}/°C, and that of the polyimide is 50×10^{-6}/°C. The difference of these expansion coefficients was large, but the FGM was not cracked. The polyimide and Cu are ductile and can relax residual stress by plastic deformation. A small disk FGM of 20 mm was not deformed, but a large disk of 50 mm disk was curved by the large shrinkage of the polyimide, as shown in Figure 3.10. The bonding strength between the polyimide and Cu was tight enough not to be ruptured.

1.2.5 WC/Mo or WC/WC(Co) FGM

The production of a WC compact is difficult without sintering additives. WC-Co cemented carbide is valuable for use in cutting tools and dies. The hardness of WC-Co systems is degraded by increasing the addition of Co [33], and their performance is below what might be expected. There are two methods for increasing the hardness of WC-Co hard metals: surface coating and reducing the amount of the Co additive. A hard metal coated with TiC and TiN is available, but a WC-Co material with less Co added has not yet become available for commercial use.

Table 3.2. Anisotropic toughness and Vickers hardness of WC/Mo FGM.

	Indentation point				
	1	2	3	4	5
Periphery					
Kc ($MPa \cdot m^{1/2}$)	4.5±0.9	7.1±0.6	7.3±1.0	7.9±1.2	8.6±1.0
Hv (GPa)	24.3±0.8	25.3±1.0	25.1±0.8	25.5±1.0	25.3±1.1
Radius					
Kc ($MPa \cdot m^{1/2}$)	7.5±0.8	7.4±0.9	7.6±1.0	8.2±0.9	8.6±1.0
Hv (GPa)	24.3±0.8	25.0±0.7	24.8±0.7	25.0±1.3	25.3±1.1

1.2.5.1 WC/Mo FGM

The thermal expansion coefficients of WC and Mo were 5.4×10^{-6}/°C and 6.9×10^{-6}/°C, respectively. There was a solid solution indicated by the chemical composition of $Mo_{1-x}W_xC$. The thermal expansion coefficient of MoWC was 6.7×10^{-6}/°C. Two layers of 70vol%WC-30vol%Mo and 30vol% WC-70vol%Mo were inserted between the WC and Mo layers. The inserted powders reacted and became the mixtures of WC-$Mo_{1-x}W_xC$ and $Mo_{1-x}W_xC$ - Mo. The expansion of the mixed composition was close to that of Mo. The two inserted layers guarded against the cracking of the FGM, but did not completely relax the residual stress. The pure WC compact has rather high toughness of 6.0 MPa compared with other ceramics [25]. The fracture toughness and Vickers hardness of the WC/Mo FGM are shown in Table 3.2.

These values were obtained by Vickers indentation. The disk sample was 2 cm in diameter. An indenter was impressed at five points on the disk, as shown in Figure 3.11. The crack lengths from the corners of the indentation were anisotropic, they extended longer along the circumference than along the radius, as shown in Figure 3.12. Toughness was separately measured from the crack lengths along the circumference and radius. At the same time the Vickers hardness was calculated by the same method. Vickers hardness was not anisotropic, and was the same everywhere. The periphery toughness calculated from the crack along the circumference was lower than the radial toughness [32]. Tensile stress is induced in the WC layer by the compressive stress of the Mo layer, and degrades the toughness near the edge of the disk. Even though the amount of the residual stress is not large, it reduces the toughness of the hard WC surface.

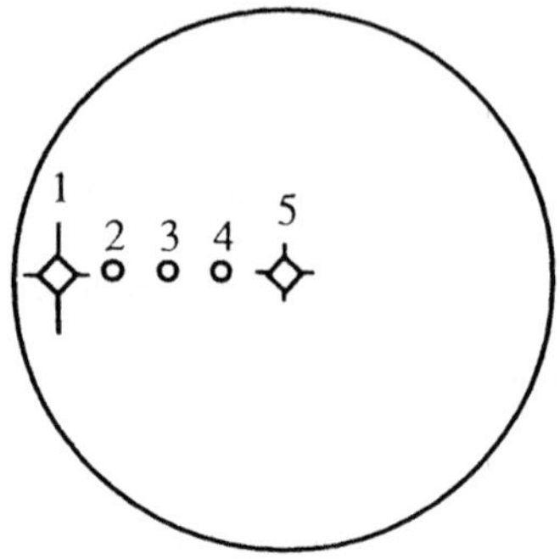

Figure 3.11. Indentation points on an FGM.

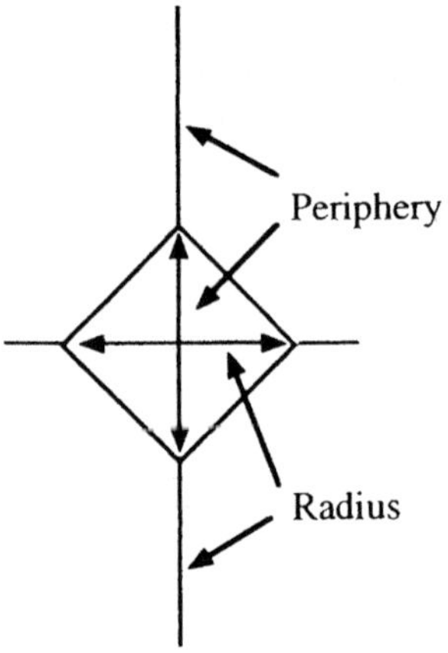

Figure 3.12. Anisotropic crack length.

1.2.5.2 WC/WC(Co) FGM

The WC disk was joined with the WC-Co(10wt%) powder. The Co element penetrated slightly into the disk, even though the sintering time was a short 5 min. A thin Co-rich layer was observed at the interface of the WC and WC-Co layers. The strength of this thin layer may be weak enough to be fractured by a large stress. It is necessary to bond the WC disk and the WC-Co layer. The Co element may diffuse into the WC layer by heating for a longer time such as for 30 min. The toughness and Vickers hardness of the FGM were measured, as shown in Table 3.3. Vickers hardness decreased slightly at the edge, because the eutectic composite of the Co and WC diffused through small spaces between the sample and a die wall. Toughness did not decrease near the edge of the disk sample. The thermal expansion of the WC-Co composite was almost same as that of the WC. Residual stress was not induced between the WC and WC-Co layers, and did not degrade the toughness of the WC layer.

Table 3.3. Anisotropic toughness and Vickers hardness of WC/WC (Co) FGM.

	Indentation point				
	1	2	3	4	5
Periphery					
Kc ($MPa{\cdot}m^{1/2}$)	6.7±0.8	7.2±0.3	7.5±0.4	7.4±0.3	7.6±0.4
Hv (GPa)	22.2±0.4	24.0±0.3	25.3±0.2	24.6±0.4	25.0±0.3
Radius					
Kc ($MPa{\cdot}m^{1/2}$)	6.2±1.1	7.7±0.4	8.1±0.6	8.2±0.5	8.6±0.6
Hv (GPa)	22.8±0.9	24.1±0.4	24.8±0.3	25.0±0.7	25.3±0.6

1.2.6 Summary

Thermosetting polyimides and WC powders are not consolidated by heating. Spark plasma sintering (SPS) makes it possible to prepare dense compacts from these two powders.

(1) A dense polyimide/Cu FGM was fabricated from thermosetting polyimide and Cu powders by SPS. The residual stress due to the thermal expansion mismatch of the polyimide and Cu was relaxed by their plastic deformation.

(2) The WC/Mo FGM with two inserted layers did not have pores or cracks. The thermal expansions of the WC and Mo are different. Tensile stress was created on the WC layer by this difference. The fracture toughness of the surface of the WC layer was degraded near the edge of the disk sample by the tensile stress.

(3) The residual stress of the WC/WC (Co) FGM was slight, because the thermal expansions of WC and WC(Co) were almost same. The toughness of the WC layer did not decrease even near the edge of the disk sample. This FGM has a hard enough surface to be used as a cutting tool.

1.3 Advanced SPS Systems

1.3.1 Introduction

The goal of this research is to develop large-size ceramic/metal bulk functionally graded materials (FGM/LB) featuring thermal stress relaxation, high heat-resistance, high wear-resistance, and high mechanical strength for highly efficient engine component applications and production type systems for FGMs. The approach focused on the use of the Spark Plasma Sintering (SPS) method [34], a pressure-assisted pulsed current sintering technology, for FGM/LB fabrication utilizing a large experimental SPS apparatus and automatic powder stacking equipment. As a part of the development program for practical production processes and machines for FGMs using SPS, the processes, mechanical properties, dimensional size and shape effects, and production machine systems were investigated.

In the past, ZrO_2/stainless steel, ZrO_2/TiAl, ZrO_2/Ni, Al_2O_3/Ti, WC/Co, WC/Co/steel, Al/polyimide, Cu/polyimide, nano-composites, porous and other combinations of bulk FGMs have been processed using SPS. Most of the specimen sizes were small, however, in a range of 20 to 30 mm in diameter [35].

Large-size ZrO_2 (3Y)/stainless steel FGMs with diameters of 50, 80, 100 and 150 mm and thicknesses of 15 and 17 mm were homogeneously fabricated for industrial use on the recently developed and world's largest Spark Plasma Sintering (SPS) System and a fully automated FGM/LB Manufacturing System [36].

1.3.2 The Spark Plasma Sintering (SPS) Process

Spark Plasma Sintering (SPS) is a newly developed process—a synthesis and processing technique—which makes possible sintering and sinter-bonding at low temperatures and in short periods by charging the intervals between the powder particles with electrical energy and effectively applying a high temperature spark plasma generated momentarily at the beginning of the pulsed current energizing. SPS systems offer many advantages over

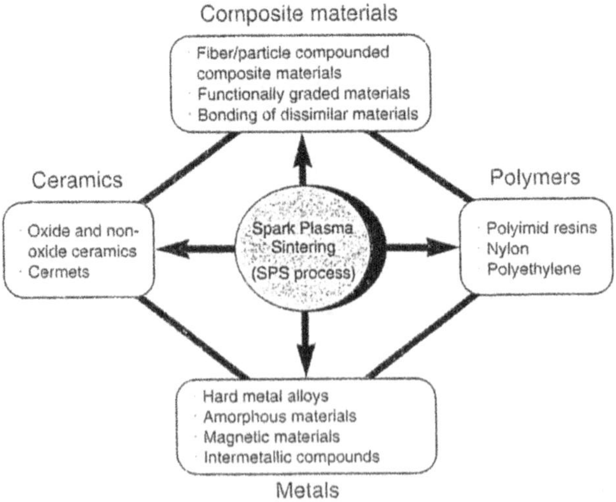

Figure 3.13. Typical examples of materials covered by SPS processing (Source: SPS Forum survey).

conventional methods using hot press (HP) sintering, hot isostatic pressing (HIP) or atmospheric furnace sintering. Figure 3.13 shows the materials covered by SPS processing.

Figures 3.14 and 3.15 show stand-alone type Automatic Powder Stacking Equipment for FGM/LB and a large-size Spark Plasma Sintering System. The sintering graphite die and punch are handled by a robot mounted on the left side of the Sintering Machine.

Figure 3.14. Automatic powder stacking.

1.3.3 Principles of the SPS Process

The SPS process features very high thermal efficiency because of the direct heating of the sintering graphite mold and the stacked powder materials by a large spark pulse current. It can easily consolidate a homogeneous, high-quality sintered compact because of the uniform heating, surface purification and activation made possible by dispersing the spark and/or Joule heating points. Figure 3.16 shows the basic configuration of a typical SPS system.

The system consists of a SPS sintering machine with a vertical single-axis pressurization mechanism, specially designed punch electrodes incorporating a water cooler, a water-cooled vacuum chamber, a vacuum air-argon gas atmosphere control mechanism, a special DC-pulse sintering power generator, a cooling water control unit, position measuring and temperature measuring units, an applied pressure display unit and various interlock safety devices.

Figure 3.15. 3,000 KN large-size spark plasma sintering system.

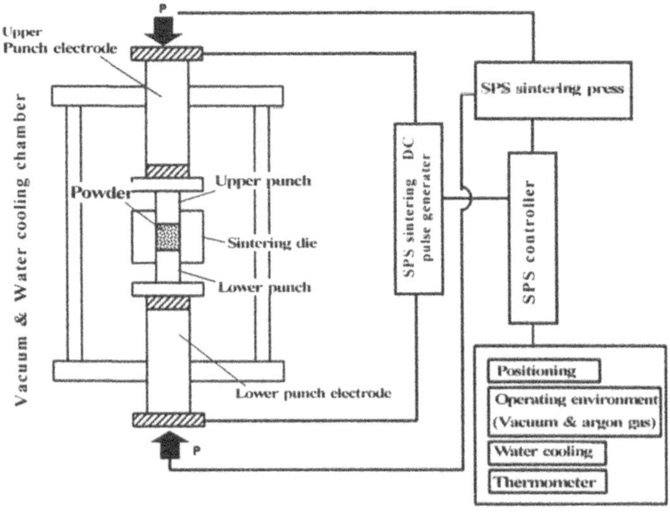

Figure 3.16. SPS system configuration.

1.3.4 Examples of Synthesized FGMs

With the SPS process, high density sintered specimens of FGMs can be fabricated at much lower temperatures and with shorter heat-up and holding times than by hot pressing.

Figure 3.17 shows typical examples of bulk FGM compacts synthesized by Spark Plasma Sintering. They are, from the left, ZrO_2 (3Y)/stainless steel with 6 interlayers, ZrO_2 (3Y)/nickel with 7 interlayers, copper/stainless steel with 5 interlayers, aluminum/polyimide with 3 interlayers, and Al_2O_3/titanium with 3 interlayers to the right. No pores or cracks were detected on the disk-shaped sintered compacts obtained by this process.

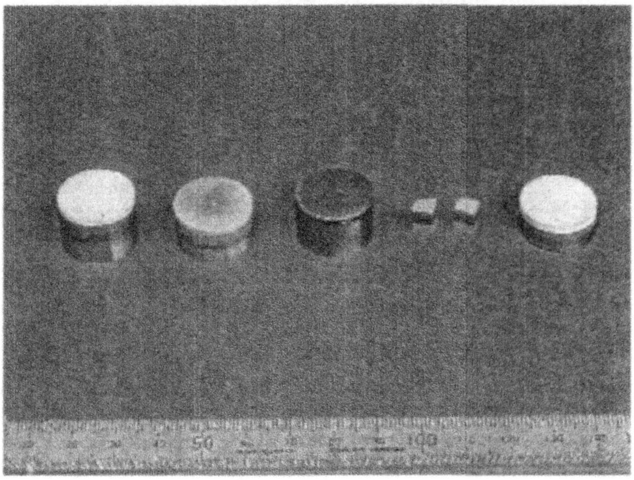

Figure 3.17. Typical example of bulk FGM compacts fabricated by SPS (From the left: ZrO_2/SUS, ZrO_2/Ni, Cu/SUS, Al/Polyimide and Al_2O_3/Ti).

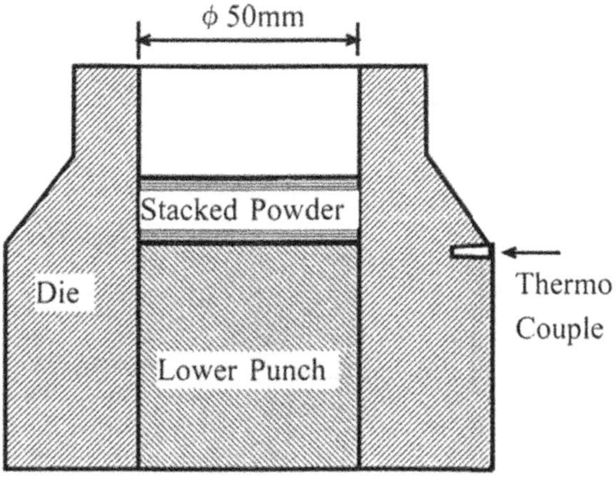

Figure 3.18. Graphite temperature gradient die/punch.

1.3.4.1 ϕ50 mm ZrO_2(3Y)/Stainless Steel FGM/LB

As shown in Figure 3.18, a 3 mol% yttrium partially stabilized zirconia (PSZ) powder, a SUS410L stainless steel (SUS) powder and their mixed powders as intermediate layers were stacked in a graphite temperature gradient die with a 50 mm internal diameter. Table 3.4 shows the layer composition and thickness of each layer after sintering. The SUS powder has an average particle size of 9 μm while the PSZ powder is composed of granulated particles with an average particle size of 50 μm (its crystalline size is 350 μm). The sintering pressures used were from 20 to 40 MPa, at SPS temperatures of 1243 or 1293K, with a temperature rise rate of 50K/min. The temperature was measured near the SUS layer.

Table 3.4. Composition and thickness of FGM layers.

Layer No.	Layer Composition(vol%)		Thickness After Sintering (mm)
	SUS410L	PSZ	
1	0	100	3
2	10	90	1
3	20	80	1
4	30	70	1
5	40	60	1
6	50	50	1
7	60	40	1
8	70	30	1
9	80	20	1
10	90	10	1
11	100	0	3

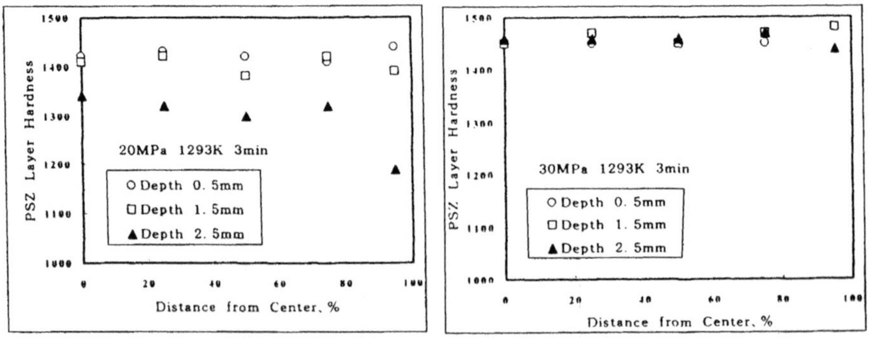

Figure 3.19. Relationships between pressure and PSZ layer hardness distribution at 1293K.

Figure 3.19 shows the relationships between the sintering pressures at SPS temperatures of 1293K and the hardness distribution in the PSZ layer. The FGM specimens sintered with a sintering pressure of 20 MPa had degraded hardness in the lower parts of the PSZ layers, with the hardness being most degraded in the areas near the perimeter. By contrast, the FGM specimens sintered at pressures of 30 MPa or higher showed hardly any difference in hardness either by depth or diameter.

1.3.4.2 φ150 mm FGM/LB

The starting materials in this study were ZrO_2 (3Y) submicron-size partially stabilized zirconia powder containing 3 mol % Y_2O_3 and SUS410L stainless steel powder with an average particle size of 9 μm and 60 μm. The composition of the stainless steel was 0.03 wt% C, 0.89 wt% Si, 0.13 wt% Mn, 0.011 wt% P, 0.012 wt% S, 0.09 wt% Ni and 13.06 wt% Cr. A total of 11 gradient layers were laminated by stacking nine kinds of mixed-composition powders with ZrO_2 (3Y)/stainless steel ratios of 90/10, 80/20, 70/30, 60/40, 50/50, 40/60, 30/70, 20/80 and 10/90 vol. % as interlayers between the 100% front and back layers. As a result of optimizing SPS conditions, it is now possible with the SPS method to fabricate within an one hour, including heat-up time, holding time and cooling time, a fine ZrO_2(3Y)/stainless steel large-size FGM/LB with a diameter of 150 mm, thickness of 15 mm, and 9 interlayers that are free from cracks, distortion and delamination as shown in Figure 3.20.

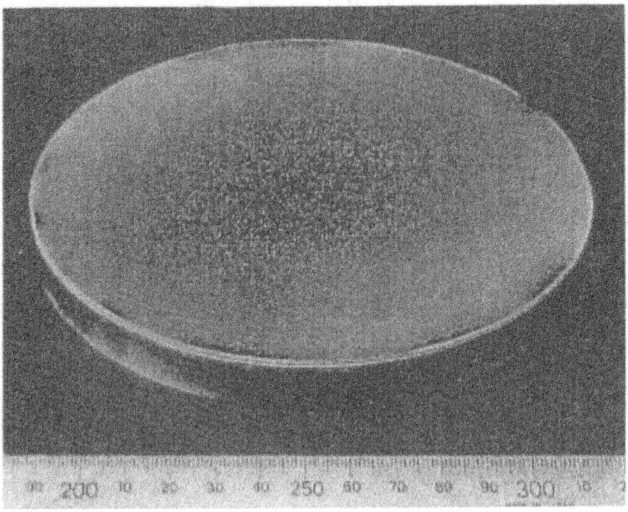

Figure 3.20. φ150 mm ZrO_2(3Y)/stainless steel (SUS410L) FGM/LB sintered compact.

1.3.5 Conclusion

A large-size ZrO_2(3Y)/stainless steel FGM/LB was homogeneously consolidated on a fully automated SPS systems. There is, however, still a variety of R&D issues to be addressed before the full potential of the SPS process for FGM/LB is realized. These issues include systems automation, increasing the pulse current capacity and versatility of functions of the hardware, and the creation of a sintering technology database and software improvements for reliability and reproducibility.

1.4 Automatic FGM Manufacturing System

1.4.1 Introduction

From 1997 to 1998, a basic sintering process was developed to fabricate large-size ceramic/metal functionally graded bulk materials (FGM/LB) with thermal stress relaxation and heat-resistant properties using an experimental large-size SPS apparatus. An automatic materials handling robot system and an independent FGM powder stacking system used with the spark plasma sintering (SPS) process were also developed. Based on the optimized sintering conditions and device development technologies acquired through the above research, the world's first tunnel type automatic FGM/LB manufacturing system was developed in 1999. The system targets the optimization and practical implementation of the sintering process for large FGMs with diameters of ϕ 100 and ϕ 150 mm. Although the fabrication of FGMs usually takes many hours, the development of this new automated SPS system now makes it possible to process such materials in a significantly shorter period of time, within 1 hour or less, by using features inherent in SPS quick sintering technology.

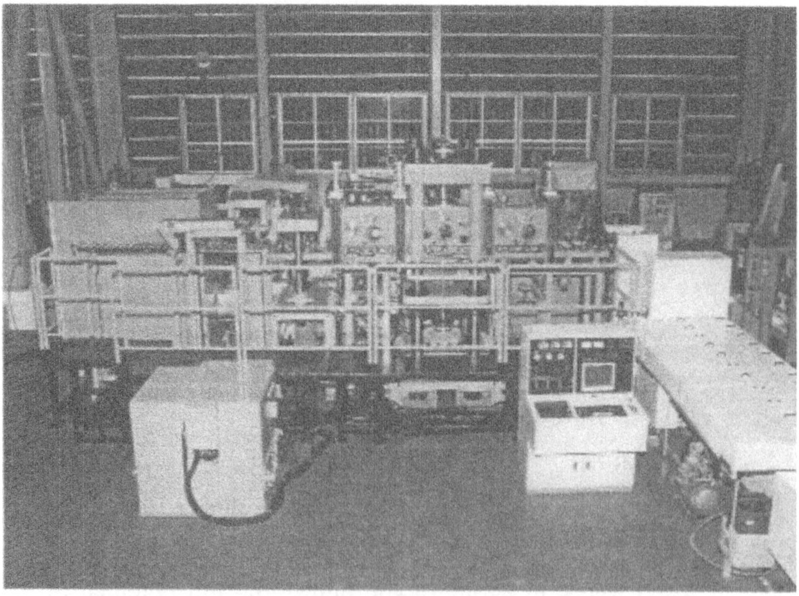

Figure 3.21. External view of the fully automated FGM/LB manufacturing system.

1.4.2 System Configuration

The fully automated FGM/LB manufacturing system is composed of a DC pulse power supply, the FGM/LB manufacturing unit itself and their hydraulic units and control panels. The FGM/LB manufacturing unit is composed of an automatic FGM powder stacking stage, a tunnel type SPS system preheating stage, SPS sintering and FGM cooling stages, an automatic die release stage, an automatic material conveyer system and an automated control system. All of these components are modularized on a per-stage unit basis. Figure 3.21 shows the fully automated FGM/LB manufacturing system with safety covers removed.

1.4.2.1 Automatic FGM Powder Stacking Stage

The powder stacking system is designed to be compatible with any future increase in container throughput by arranging them in series. The stacking powder is selected with a self-driven cart movement system in which the cart carrying the sintering dies/punches travels below the containers.

The method used for the powder stacking is the volume method (to only fill the container), and the modular configuration uses material powder containers of different volume fractions arranged in a line to increase or decrease the number of containers depending on the number of stacked layers.

Following powder stacking, the FGM sintering dies/punches made of graphite are carried by an elevator up to the height of the conveyer line of the preheating stage. They are then carried together with the carrying tray to the roller drive section at the inlet of the atmosphere control stage. Table 3.5 shows the main specifications of the automatic FGM powder stacking stage.

Table 3.5. Main specifications of automated FGM powder stacking stage.

Item	Specification
Powder stacking method	Cart-based, volumetric method
Powder weight measurement	Load cells
Container movement	Linear motor drive
Powder dispersal prevention	Retainer absorption
Number of powder containers	11 (11 layers)
Successive packing count	5
Powder stacking	500 kN

1.4.2.2 Tunnel Type SPS System (Atmosphere Control/SPS Sintering/FGM Cooling Stages)

Each of the chambers of the atmosphere control, SPS sintering and FGM cooling stages has an independent vacuum exhaust system and gas inlet system, so that these stages can be used independently in separate inactive gas atmospheres. Gate valves are installed between the chambers, at the inlet of the preheating chamber and at the outlet of the FGM cooling chambers. Table 3.6 shows the main specifications of the preheating, SPS sintering and FGM cooling stages.

Table 3.6. Main specifications of preheating, SPS sintering and FGM cooling stages.

Item	Specification
Pressurizing method	Hydraulic cylinder raising
Maximum pressure	2 MN
Pressure control range	20 kN to 2 MN
Pressure measurement	Load cells
Degree of vacuum	6 Pa
Sintering stage type	Water-cooled jacket cooling
Cooling stage type	Water-cooled jacket water cooling
Cooling method	Inactive gas introduction

(1) Preheating Stage

The preheating stage has two roles; one is the preheating adjustment chamber for the sintering chamber that regulates and equalizes the pressures of the SPS sintering and atmospheric control chambers, opens the gate valve between the chambers and feeds the sintering die/punch (carrying tray). The other is accepting the carrying trays from the FGM powder stacking stage.

(2) SPS Sintering Stage

The SPS sintering stage performs the SPS sintering of the powder stacked in the FGM sintering die/punch by means of pressurization and a pulse power supply. This corresponds to the sintering machine in conventional SPS systems. The extremities of the electrodes are integrated with graphite spacers. The window for observation of the SPS has a 150 mm, large-diameter wiper/shutter mechanism and is attached to the main body using a flange system for easy maintenance.

(3) FGM Cooling Stage

The FGM cooling stage moves the sintering die/punch after SPS from the SPS sintering chamber to the FGM cooling chamber and then cools the temperature-graded graphite material and FGM to a temperature at which the material can be taken out into the atmosphere. The sintering die/punch and sintered compact can be cooled either by argon gas or by heat transfer in which a material of high thermal conductivity such as copper or aluminum is brought into contact with the sintering die/punch.

1.4.2.3 Automatic Die Release Stage

At the position where the sintering die/punch and FGM sintered compact are separated, the carrying tray is stopped by the home position stopper, then a push rod driven by a hydraulic cylinder is raised from underneath to separate the sintering die/punch and the FGM (sintered compact). The FGM is pushed up from the sintering die/punch and is clamped by a manually operated air cylinder and placed on the slat conveyer. An FGM compact sensor is installed at the conveyer outlet. When an FGM sintered compact is detected, it issues an alarm so that subsequent FGMs are not conveyed.

1.4.2.4 Automatic Material Conveyer System

The FGM powder compact and the temperature-graded sintering die/punch are placed on special SPS carrying trays, and sent from the sintering die/punch stockpile to the FGM powder stacking stage, press unit and finally to the tunnel type SPS sintering stage. After a pulse power supply has been applied for a specified time period, the tray is moved to the cooling stage, and then to the die release stage where the die/punch and FGM sintered compact are separated and placed on their respective conveyers.

The sintering die/punch is returned to the sintering die/punch stockpile for reuse.

1.4.2.5 Automatic Control System

The automatic control system comprises two personal computers, one working as the host managing the system and the other processing data for data management. Two sequencers including the main one that controls the main system (atmosphere control, SPS sintering and FGM cooling stages) and the slave sequencer that controls the FGM powder stacking stage are also included.

The operator-machine interface for inputting setting data, issuing operational commands and monitoring the operational conditions is a large touch panel integrated with the host personal computer's monitor.

1.4.2.6 Safety System Devices and Other Peripherals

The safety system devices enable monitoring from the control panel of the cooling water temperatures in the SPS sintering chamber, FGM cooling chamber, and the upper/lower electrodes, hydraulic/vacuum units and the conveyer of the SPS sintering stage.

To minimize the effects in the event a die or punch is damaged during the SPS process, a die/punch damage detector system monitors a load cell signal caused by the pressure thrust using a sequencer. This system detects abnormal changes in the signal as a sign of die/punch damage, and stops the power supply and pressurization.

1.4.3 FGM/LB Sintering by the Fully Automated SPS System

A ZrO_2 (3Y)/stainless steel (SUS410L) FGM/LB with a diameter of 150 mm was sintered using the fully automated SPS system and obtained the results described in the following.

Table 3.7. FGM/LB sintering conditions of the fully automated FGM/LB SPS manufacturing system.

Sintering Conditions	
Sintering die/punch	Temperature graded sintering die/punch made of graphite
Atmosphere (Pa)	2.0
Pressure (MPa)	12
Sintering temperature (K)	1213
Temperature heating-up time (min.)	50
Holding time (min.)	4
Cooling time (min.) (Atmospheric exposure time of the cooling stage)	50

The powder was stacked in a special die for the fully automated SPS system by setting the stacking quantities at the powder stacking stage. The thickness of the inter-mediate layers of the φ 150 mm FGM/LB sintered compact was 1 mm, and the thickness of the ZrO_2 (3Y) and stainless steel (SUS410L) was 3 mm. To check the condition of the stacked layers, preparatory sintering of the φ150 mm ZrO_2 (3Y)/stainless steel (SUS410L) FGM/LB was performed and the sintered compact was inspected by cutting it. The layer boundaries were found to be relatively satisfactory except for some disturbance at the interface between the pure stainless steel powder and the layer of the 10/90 vol % mixture of ZrO_2 (3Y) powder/stainless steel (SUS410L) powder. Figure 3.22 shows the condition of the layers. After confirming the laminating condition, the φ 150 mm ZrO_2 (3Y)/stainless steel (SUS410L) FGM/LB was sintered with the fully automated FGM system using the sintering conditions shown in Table 3.7. A FGM/LB sintered compact, without any cracks or distortion was successfully sintered in less than an hour as shown in Figure 3.23.

Figure 3.22. Laminating condition of φ150 mm ZrO_2 (3Y)/stainless steel (SUS410L) FGM/LB.

Figure 3.23. φ 150 mm ZrO_2 (3Y)/stainless steel (SUS410L) FGM/LB sintered compact.

1.4.4 Conclusions

In order to develop large-size ceramic/metal bulk FGMs (FGM/LB) using the spark plasma sintering (SPS) method, the multi-layer composition,

optimum sintering conditions and the manufacturing machine systems were investigated. The study resulted in:
(1) Development of the world's first fully automated FGM/LB manufacturing system that targets the optimization and practical implementation of the sintering process for FGM/LB with diameters of ϕ 100 and ϕ 150 mm.
(2) This automated SPS system made possible the automatic fabrication of a 150 mm diameter, 15 mm thick ZrO_2 (3Y)/stainless steel FG/LB with 11 gradient layers in 54 minutes (50 minutes heat-up time, 4 minutes holding time and 50 minutes cooling time).

To meet practical industrial needs such as complex shapes and properties including 3-dimensional gradient FGM components, however, further development of automatic SPS manufacturing systems is required.

2. COATING PROCESS

2.1 Plasma-Sprayed Thermal Barrier Coating

2.1.1 Introduction

A major failure mechanism in plasma-sprayed thermal barrier coatings (TBCs) is spallation of the top coat due to the top/bond coat thermal expansion mismatch concomitant with deposition-induced interfacial roughness and oxide film growth. Preliminary experiments suggest that reduction of this mismatch by embedding alumina particles into the bond coat may increase TBC durability [37]. This approach is examined using the higher-order theory for functionally graded materials [38], by investigating the effect of a graded bond coat microstructure on the thermally-induced through-thickness stress in the vicinity of a rough top/bond coat interface's crest. This stress component initiates horizontal cracks in the porous plasma-sprayed top coat at or just above the rough interface's crest whose propagation is facilitated by the growth of an interfacial oxide film. The analysis is conducted in two distinct ways which reflect the present dichotomy in the modeling of functionally graded materials. The coupled approach explicitly accounts for the micro-macrostructural interaction and is an intrinsic feature of the higher-order theory. The uncoupled approach, which is the prevailing paradigm in the mechanics/materials communities, replaces the actual graded microstructure with a fictitious material having continuously or discretely varying properties. These fictitious properties are then used in the structural analysis of a TBC without regard for particle-particle or particle-interface interactions. The limitations of this approach are discussed vis-a-vis managing stress fields in TBCs with functionally graded bond coats.

2.1.2 Higher-Order Theory

The higher-order theory provides an approximate framework (demonstrated to be sufficiently accurate by comparison with finite-element benchmark results) for calculating temperature and displacement fields in materials/structural components with spatially varying microstructures. The analytical approach employed in the theory's construction explicitly couples the micro-macrostructural interaction, in contrast with the standard uncoupled micromechanics approach based on the representative volume element concept employed by the majority of researchers in the analysis of FGMs, which may produce unacceptably inaccurate results. An important advantage of the higher-order theory over the standard displacement-based finite-element analysis is the simultaneous satisfaction of displacement and traction continuity conditions between different regions of the spatially variable microstructure, which is important for accurately modeling rapidly changing field variables across material discontinuities and spatially variable geometric details.

In order to model functionally graded materials by the higher-order theory, the volume occupied by the material is discretized into a number of generic cells whose dimensions and material parameters are appropriately assigned to reflect a given microstructure's geometry and spatial variation. As demonstrated in previous studies on functionally graded TBCs with different microstructural scales [39], and duplex TBCs with different degrees of interfacial roughness and oxide film thickness [40], the volume discretization employed in the higher-order theory allows sufficiently detailed representation of spatially variable material architectures and geometry. For a given thermomechanical loading, an approximate solution for the temperature and displacement fields in each cell is constructed using a higher-order expansion in the local coordinates in order to capture the local effects created by the thermomechanical field gradients and the material's functionally graded microstructure. The unknown coefficients in the temperature and displacement field expansions are obtained by constructing systems of equations that satisfy the boundary-value problem requirements for the given field variable approximations. That is, zeroth, first and second moments of the heat and equilibrium equations are satisfied in a volumetric sense, and the thermal and heat flux continuity conditions, and the displacement and traction continuity conditions, within a given cell, as well as between a given cell and its adjacent neighbors, are imposed in an average sense, together with the thermomechanical boundary conditions (see Ref. [38] for details).

2.1.3 Model Geometry

The model geometry of the investigated TBC is shown in Figure 3.24. The rough top/bond coat interface is modeled using a sinusoidal waveform, as is customary. The top coat is porous zirconia, while the bond coat and the substrate are NiCr-based and Ni-based alloys, respectively. The alumina particles embedded in the bond coat are limited to the immediate vicinity of the top/bond coat interface as will be discussed in more detail subsequently. The inelastic response of these materials is modeled using the power-law creep model generalized to multiaxial loading. The thermoelastic and creep material parameters of these constituents have been reported in Ref. [40]. For a spatially uniform thermal cycle from 25 to 1150 to 25 deg. C of 630 sec duration (with approximately a 220 sec exposure at 1150 deg. C), which simulates a furnace durability test, the evolution of the through-thickness stress σ_{22} at the critical crest and trough locations of the wavy interface throughout the thermal loading history has been reported in Ref. [40] for a TBC with a pure NiCr-based bond coat without the alumina particles. Due to the stress relaxation at the elevated temperature, the through-thickness stress σ_{22} at the wavy interface's crest becomes tensile upon cooldown, thereby promoting horizontal crack initiation in the weak porous top coat. It has been suggested in Ref. [37] that the magnitude of this stress component at the wavy interface's crest can be reduced by dispersing submicron size alumina particles in the bond coat close the interface. This effectively reduces the thermal expansion coefficient mismatch between the top and bond coats, albeit at the expense of increasing the effective moduli.

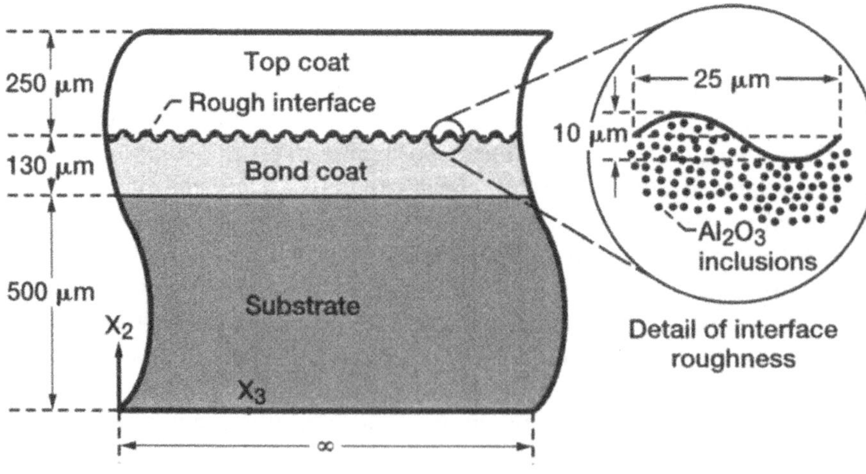

Figure 3.24. Model geometry of the investigated graded bond coat TBC.

Preliminary investigation into the reduction of the through-thickness interfacial stress due to uniform dispersions of alumina particles in the bond coat, based on the higher-order theory analysis with the actual heterogeneous bond coat microstructure replaced by equivalent properties calculated using the Generalized Method of Cells (GMC) micromechanics model [41], indicated a continuous decrease of this stress at the wavy interface's crest with increasing alumina particle content. However, this was accompanied by an increase in the magnitudes of the compressive through-thickness and transverse stresses, σ_{22} and σ_{33}, respectively, as well as the inplane shear stress σ_{23}, in the trough region due to the increased effective moduli of the heterogeneous bond coat. Therefore, the interfacial bond coat region was graded from a high alumina particle content of 40% at the crest to 0% below the trough in order to manage the stresses at both locations. A detail of the interfacial region showing the discretization employed in the higher-order analysis, and the graded alumina particle distribution, is shown in Figure 3.25(*l*). The size of the square alumina particles is 0.5 μm which corresponds to the smallest square cell dimensions used to discretize the interfacial region. The results generated using this microstructure are compared in the sequel with the results obtained using the homogenized interfacial region microstructure shown in Figure 3.25(*r*). This micro-structure was generated from the actual heterogeneous microstructure by subdividing the graded region into 11 horizontal layers and then calculating the effective thermoelastic and creep parameters in each layer (based on the local alumina

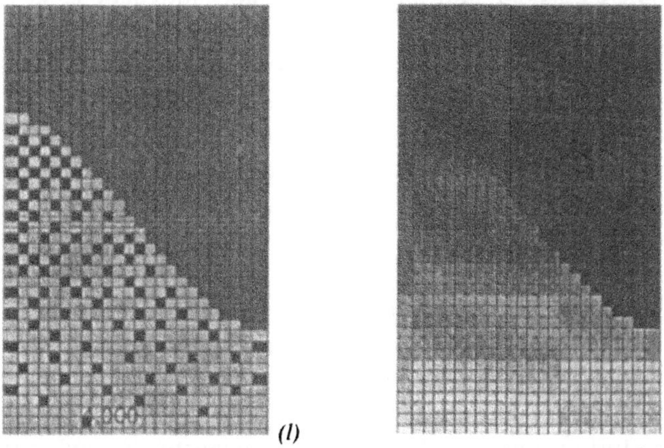
(*l*) (*r*)

Figure 3.25. Detail of the volume discretization in the interfacial region: (*l*) heterogeneous Microstructure ; (*r*) homogenized microstructure.

particle content) using GMC. The model shown in Figure 3.25(*l*) explicitly accounts for the particle-particle and particle-rough interface interaction that is absent in the homogenized treatment of the functionally graded microstructure shown in Figure 3.25(*r*), which is the current practice in the mechanics/materials communities.

2.1.4 Results and Discussion

Results for the through-thickness and transverse stress distributions σ_{22} and σ_{33} along the vertical cross section passing through the crest of the top/bond coat interface, obtained at the end of the thermal cycle from the homogenized and fully coupled analyses, are shown in Figure 3.26. Included in the figure are the corresponding distributions obtained for the TBC with a pure NiCr-based bond coat without alumina particles. Of particular interest is the magnitude of the through-thickness stress σ_{22} in the top coat at or just above the top/bond coat interface, shown in Figure 3.26(*l*), i.e, for $x_2 \geq 635$ μm, as this is the horizontal crack initiation site. The higher-order theory analysis based on the homogenized properties of the graded bond coat region predicts a substantial reduction of this stress component at the top/bond coat interface relative to the TBC without the alumina particles embedded in the bond coat, i.e., 48 MPa versus 76 MPa. This indicates that the extra cost involved in processing TBCs with heterogeneous bond coat microstructures in the interfacial region would be mitigated by a potentially high benefit derived from the thermal expansion mismatch reduction due to the alumina particles. However, the reduction in the through-thickness stress is rather modest when the fully coupled analysis, which explicitly accounts for the particle-particle and particle-interface interaction, is employed. Pindera *et al.* note that the differences between the homogenized and fully coupled analyses, which are naturally expected in the heterogeneous region, also occur in the homogenous top coat region. These differences appear to results from the particle-interface interaction which is absent in the homogenized analysis. This has important consequences regarding the use of heterogeneous bond coats to reduce the through-thickness stress σ_{22} at the top/bond coat interface through reduction of the thermal expansion mismatch.

A related important point regarding the particle-interface interaction taken into account in the fully coupled analysis is the size effect of the alumina particles. It is conceivable that the extent of this interaction may diminish with decreasing alumina particle dimensions, in which case the results of the fully coupled analysis would be expected to tend to those of the homogenized analysis. This is yet to be established. If this were the case, then the fully coupled higher-order theory analysis would provide important information on the particle size necessary to obtain substantial reduction in the through-thickness stress in the porous top coat at the horizontal crack initiation site.

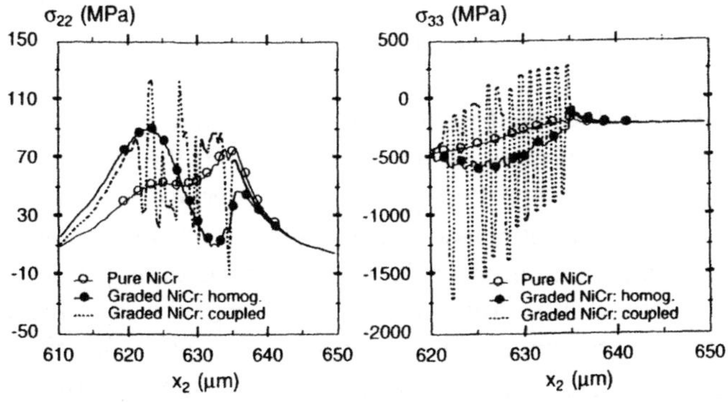

(*l*) (*r*)

Figure 3.26. Through-crest σ_{22} (*l*) and σ_{33} (*r*) distributions at t = 630 sec from fully coupled and homogenized higher-order theory analysis.

Examination of the transverse stress σ_{33} distribution shown in Figure 3.26(*r*) leads to the conclusion that the homogenized analysis provides an average distribution in the functionally graded bond coat region about which the results of the fully coupled analysis oscillate. The average distribution is below that which arises in the TBC with a pure NiCr-based bond coat. However, tensile transverse stresses are predicted by the fully coupled analysis in the upper functionally graded region in the metallic phase while the homogenized analysis predicts only compressive transverse stresses in both ceramic and metallic phases. The tensile transverse stresses in the metallic phase cannot be recovered by using the results of the homogenized analysis as input in a micromechanics scheme to estimate the magnitude of the local microstresses. Finally, both analyses predict compressive transverse stress in the top coat at the top/bond coat interface, with the fully coupled analysis predicting a higher magnitude of this stress component than the homogenized analysis, i.e., –88 MPa versus –66 MPa, as was the case for the through-thickness stress at this location. Both these values are lower in magnitude than the corresponding value obtained for the TBC with a pure NiCr-based bond coat.

2.1.5 Conclusions and Future Modeling Issues

The presented results generated using the higher-order theory for FGMs indicate that micro-macrostructural coupling which explicitly accounts for particle-particle and particle-interface interaction is important in the analysis of plasma-sprayed TBCs with heterogeneous bond coat microstructures. In particular, homogenization-based analysis overestimated the extent of the through-thickness stress reduction at the crest of the rough top/bond coat interface relative to the fully coupled higher-order theory analysis for the chosen alumina particle dimensions. This stress component plays an important role in the initiation of horizontal cracks in the porous top coat, which ultimately result in spallation due to subsequent growth of an interfacial oxide film.

The approach described herein can also be employed to investigate the effect of alumina particle dimensions on the through-thickness stress evolution in the vicinity of the top/bond coat interface. It is conceivable that decreasing the particle dimensions may produce results that approach those of the homogenization-based analysis, thereby rendering graded bond coat architectures potentially attractive in reducing the interfacial through-thickness stress. This approach to increasing the TBC durability must take into account the effects of the interfacial oxide film growth/thickness [42] and the top coat porosity. Both effects can be investigated by the higher-order theory.

2.2 Thermally Sprayed FGM and Its Property

2.2.1 Introduction

Functionally graded materials are spatial composites which display discrete or continuously varying compositions over definable geometrical length. The gradients can be continuous on a microscopic level or layers comprised of metals, ceramics and polymers. The material gradients induced by spatial variations of properties make FGMs behave differently from common homogeneous materials and traditional composites. Due to this behavior, a quantification of FGM parameters is more difficult. Recently, several indentation methods were introduced to measure some of FGM properties. Suresh and coworkers [43-45] have developed analytical and experimental tools to estimate the elastic properties of graded materials using instrumented micro-indentation.

The characterization of *nonlinear* or *elastic-plastic* FGMs is more complex due to additional material parameters involved. Unlike elastic FGMs, not only the elastic modulus but also the yield stress and the strain hardening which vary through location must be determined. In order to overcome the difficulties associated with testing FGMs, Nakamura *et al.* propose a new procedure which requires a minimal effort to determine the unknown parameters. This procedure is based on the inverse analysis and makes an effective use of micro-indentation data. Specifically, the Kalman filter theory is utilized to estimate the FGM composition profile through-thickness and the stress-strain transfer parameter that defines the effective properties of elastic-plastic FGMs.

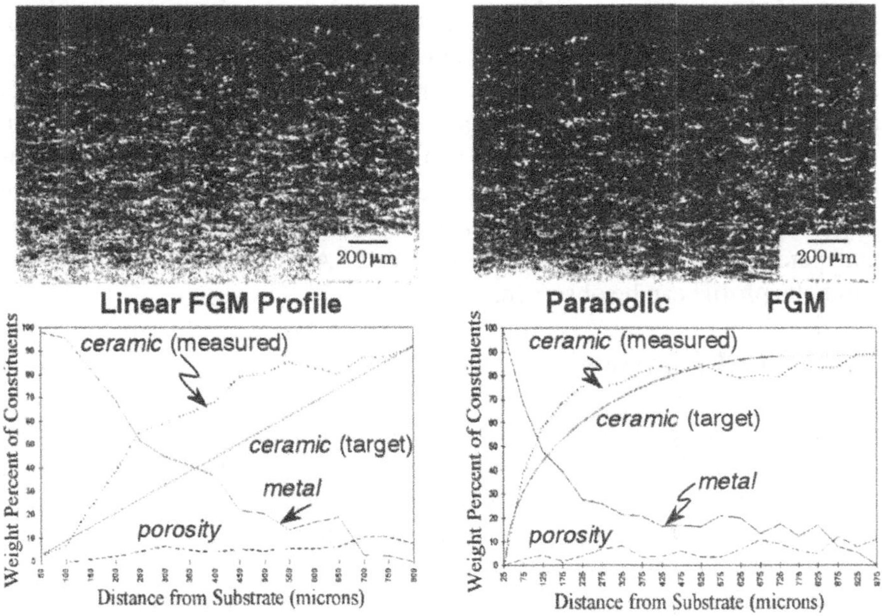

Figure 3.27. Cross-sectional micrographs and corresponding composition profiles of FGM's containing ceramic(PSZ) and metal(NiCrAlY) phases.

2.2.2 Plasma Sprayed FGM

Plasma spray processing offers a flexible and economic means for producing FGMs [46]. This method has been used for many years to apply layered and graded deposits to enhance the survivability of thick ceramic thermal barrier coatings. In plasma spraying, powder feedstock material is introduced into a plasma flame, particles melt in transit and impinge on the substrate where they flatten and undergo rapid solidification and form a deposit through successive impingement. The versatility of plasma spray allows the processing of a wide range of high performance materials, including most metals and refractory ceramics.

The gradient composing the plasma sprayed FGM is generally macroscopic (i.e., discrete layers made up of varying fractions of different constituents). The continuity of the layers is limited by the sprayed splats. Thickness of each splat is typically 1 to 10µm. By controlling feeding rates of different materials, the thermal spray can produce linear, parabolic or stepped profile grading [47]. In the case of thermal barrier coatings, it is desirable to maximize the ceramic content in the FGM to enhance the thermal insulation without sacrificing the beneficial mechanical aspects of the graded regions. In such a case, a parabolic profile would be suitable. In other applications, a linear profile may be optimal. Synthesis of desired grading profile depends on the deposition efficiency and requires integration of processing hardware, material behavior and micro characteristics. Figure 3.27 shows the FGMs consisting of yttria partially stabilized zirconia (PSZ) and NiCrAlY [47]. These ceramic and metal phases are commonly used in thermal barrier coatings for turbine and diesel engines. The compositions of the individual graded layers are measured by the X-ray diffraction and image analysis. The figure shows targeted and measured compositions for the linear and parabolic profiles of PSZ, respectively. An excellent match for the parabolic profile can be observed.

2.2.3 FGM Parameters

In an effort to establish a robust but simple procedure to determine properties of FGM, Nakamura *et al.* constructed a thin FGM model placed on top of a thick steel substrate as shown in Figure 3.28. Although there are various methods to measure mechanical responses of FGM, they consider instrumented micro-indentation whose requirements for specimen preparation are minimal. Their proposed Kalman filter process requires only gradual indentation of FGM and resulting load-displacement records.

In their finite element analysis PSZ is assumed to be linear elastic (E = 50GPa) while NiCrAlY is modeled as elastic-plastic (E = 30GPa, σ_y = 50MPa). At any location within the FGM where two phases coexist, material response is elastic-plastic. The spherical indenter is assumed to be made of a hard material (e.g., tungsten carbide). Two different radius sizes of 100μm and 500μm are considered.

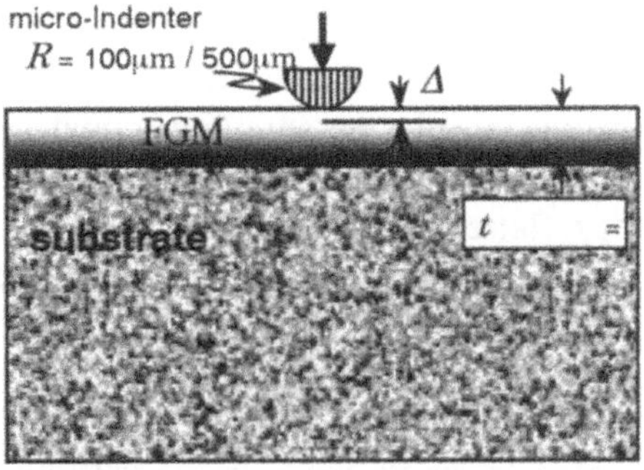

Figure 3.28. Schematic of micro-identation of FGM layer on steel substrate.

2.2.3.1 *Compositional Variation through Thickness*

As discussed earlier, one of the important FGM parameters to be determined is the composition of multiple phases through-thickness. It is assumed that the compositional variation of ceramic can be approximated by a power-law equation as, $V_c = (z\ /\ t)^n$. Here V_c is the volume fraction of ceramic phase, z is the location measured from the substrate, t is the thickness of FGM layer, and n is the power exponent. A possible range of n is assumed to be 1/3 to 3. In order to estimate this variable in the Kalman filter effectively, this power exponent is redefined as $N = \log n / 2\log 3$ ($-0.5 \le N \le 0.5$). The profiles of the compositional variation through-thickness for various N are shown in Figure 3.29.

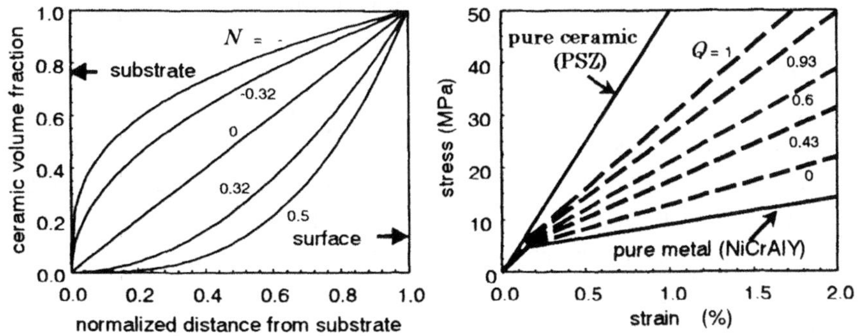

Figure 3.29. Two unknown FGM parameters to be determined in the inverse analysis; N – variation of ceramic volume fraction through-thickness, Q – stress-strain transfer parameter. Stress-strain curves are shown for pure ceramic and metal, and equal mixture composite ($V_c = 0.5$, $V_m = 0.5$) with various Q.

2.2.3.2 Effective Mechanical Properties of FGM

Another important parameter is the one which defines the effective property ceramic-metal composite. The mechanical response of FGM is dependent on the shape and contiguity, and spatial distribution of each phase. In the current analysis, so-called 'modified rule-of-mixture' described by Suresh and Mortensen (1998) is adopted. The effective properties of FGM with given V_c and $V_m = 1 - V_c$ can be determined by

$$E_f = \frac{V_c E_c + V_m E_m R}{V_c + V_m R}, \quad \sigma_{yf} = \sigma_{ym} \cdot \left(V_m + \frac{E_c}{E_m} \cdot \frac{V_c}{R} \right)$$

$$\text{where} \quad R = \frac{q+1}{q + E_m / E_c} \tag{3.1}$$

Here the stress-strain transfer parameter q plays the key role in defining the properties. The value may ranges from 0 to infinity. In the Kalman filter, it is suitably represented as $Q = \sqrt{1 - e^{-q}}$ ($0 \le Q \le 1.0$). The stress-strain relations for different Q are shown in Figure 3.29.

2.2.4 Kalman Filter

The Kalman filter was developed as an optimal recursive signal processing algorithm [48]. To estimate unknown parameters, state and covariance variables are updated at each increment with indirect measurement data (e.g., indented displacements). This method is very effective when measurements contain random error or noise since it is designed to filter them out. This method was proven to be effective in seeking unknown parameters of other materials [49].

In the current analysis, the two unknown parameters range $0 \le Q \le 1.0$ and $-0.5 \le N \le 0.5$. In the Kalman filter, $x_t = (Q, N)^T$ is set as the state vector containing these unknown components. The subscript 't' denotes the load increment (P_t) when the measured indented displacement vector Δ_t^{meas} is obtained. The measured data is assumed to contain error as $\Delta_t^{meas} = \Delta_t + \Delta_t^{err}$. At every load increment, the state vector is updated as

$$x_t = x_{t-1} + k_t\,[\Delta_t^{meas} - \Delta_t(x_{t-1})] \tag{3.2}$$

and the Kalman gain matrix k_t is calculated as

$$k_t = P_t\,\delta_t R_t^{-1} \quad where \quad P_t = P_{t-1} - P_{t-1}\delta_t^T(\delta_t P_{t-1}\delta_t^T + R_t)^{-1}\delta_t P_{t-1} \tag{3.3}$$

Here P_t and R_t are the *matrices of covariance*. The matrix δ_t contains the gradients of Δ_t with respect to the state vector. Both Δ_t and δ_t must be available for the Kalman filtering.

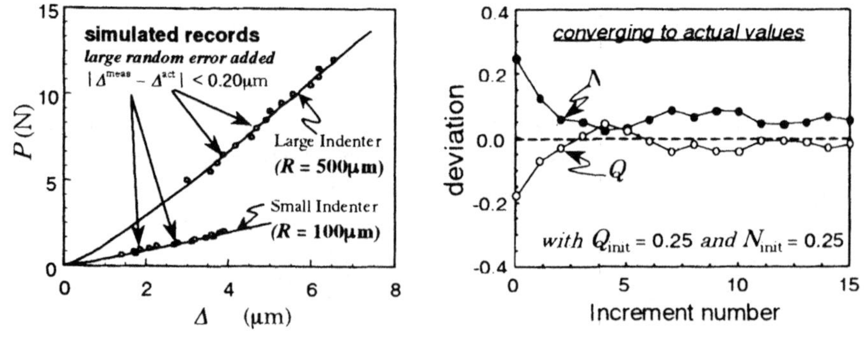

Figure 3.30. Two sets of simulated measurement records (random error added) in case of $Q = 0.43$, $N = 0$ are shown left. The right figure shows the convergence trends of two unknown parameters during the Kalman filter process. The initial estimates are $Q_{init} = 0.25$ and $N_{init} = 0.25$ while the actual values are $Q_{act} = 0.43$ and $N_{act} = 0$.

2.2.5 Simulation Study

In order to verify the current procedure, a finite element analysis is carried out. In the analysis by Nakamura *et al.*, simulated records are generated by adding large random errors (up to 0.2μm) to the finite element solutions.

For each indenter, Δ_t^{meas} at 15 load increments are generated as shown in Figure 3.30. These records are used in the Kalman filter process in (2) and (3). At beginning, the *initial estimate* of the state vector (i.e., Q and N) must be prescribed. In Figure 3.30, the filtering of one set of initial estimates is illustrated. As the increment progresses, the estimates approach the actual solutions.

Ideally, if the procedure is robust, any initial estimates should converge to the actual solutions. However, in general, some initial estimates achieve the convergence while the others do not. The size of so-called 'domain of convergence' is limited by several factors including level of measurement noise, number of available data and behavior of measured variables as functions of state vector. Initially Nakamura *et al.* have carried out the Kalman filter with the record from only one indenter (R = 100μm).

The resulting domain of convergence was small, which was an indication of poor convergence. In order to improve convergence, another indentation with a larger indenter head (R = 500μm) was combined. The result is a dramatic improvement of convergence as shown in Figure 3.31. They have also tried different values for Q and N to test the validity of this procedure and found good convergence behaviors. Based on their detailed simulation study, the Kalman filter technique has shown to be a promising and effective method to determine the FGM properties.

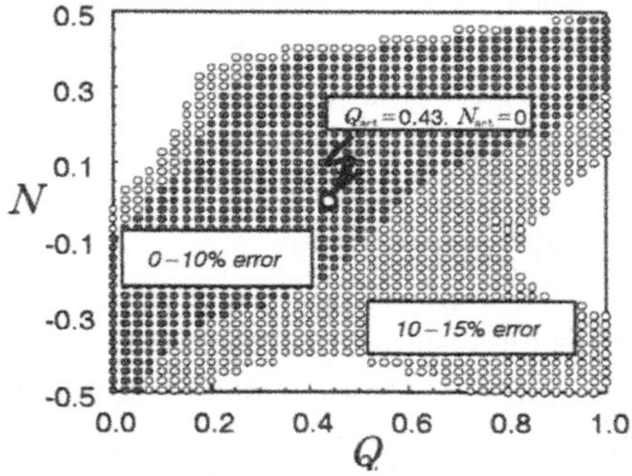

Figure 3.31. Each dot represents initial estimates that lead to convergence. A large domain of convergence supports the effectiveness of the Kalman filter.

2.3 Laser-Assisted Low-Pressure Plasma Spraying for Mo-Co Composite Material

2.3.1 Introduction

The reliability of highly developed technical systems can be increased if their various components meet the highest standards of quality control and quality assurance. To meet these highly sophisticated functional and tribological requirements, new and emerging technologies require better materials with improved properties, especially frictional characteristics. However, complex and demanding tribological requirements may be difficult to meet using conventional techniques, although conventional tribo-materials have their own industrial acclamation, which is widely recognized [50].

Amongst surface modification methods, plasma spraying is a valuable technique that makes it possible to add a coating with a new property that is present only on the sliding surface and is unrelated to the internal properties, thus creating high-performance tribo-materials. However, it has been found that there are certain practical limitations associated with this technique. These are higher levels of micro porosity, shrinkage voids and inadequate adhesion to the substrate [51]. It has been known that these limitations have significantly confined the industrial usage of this technique. Therefore, keeping in mind the industrial demands for high performance and efficient tribo materials, an attempt has been made to supplement the plasma spraying method with the addition of laser irradiation to overcome these limitations. This novel technique has proved its worth in producing a wide range of industrial coatings with favorable tribological performance [52].

Because of the practical diverse features of the molybdenum-copper alloy system, an attempt has been made to synthesize molybdenum with different percentages of copper ranging from 10 to 90 percent to create a high-performance tribo-material.

2.3.2 Preparation and Testing

The low-pressure plasma system consists of a vacuum chamber with an inside diameter of 2 m and a length of 3 m. To facilitate the spraying, the system is equipped with a 6-axis numerically controlled robot. The films were synthesized by employing a 2 mm pitch, 2000 cm/min. spray speed-coupled with a 8 mm diameter spray nozzle. A CO_2 laser rated at 1.5 KW in a continuous mode was employed using 750 amp current and 100 ton pressure during spraying. A sliding tester was used to review the tribological characteristics. For lubrication, 50-μl of engine oil was used once at the start of the test to attain boundary lubrication conditions.

2.3.3 Micro-structural Investigation

The microstructures were examined in a non-etched condition. The morphology of the microstructure plays an essential role in determining the tribological properties. The morphology depends on the solidification of the individual splats as well as the accumulation of these splats on the substrate. However, the characteristics of splats will generally depend on the nature of the material, its melting point, degree of superheating, viscosity and interactions with the substrate [53].

An enormous amount of micro porosity associated with shrinkage voids has been observed for samples with 10% copper, as shown in Figure 3.32(a) Similarly, microporosity and shrinkage voids are also prominent in the test sample prepared with 20% copper. This sample shows a significant high level of microstructure imperfection as shown in Figure 3.32(b). It has been observed that the evolution of a coating morphology is related to its thickness and spraying conditions including distance, vacuum and current [54]. Adequate adhesion with very low micro-porosity and shrinkage voids has been observed with 40% copper as shown in Figure 3.32(c). The microstructure exhibits a very homogeneous structure, along with a minimal level of microporosity, shrinkage voids, and material pull out. The adhesion to the substrate is also good and is adequate for most applications.

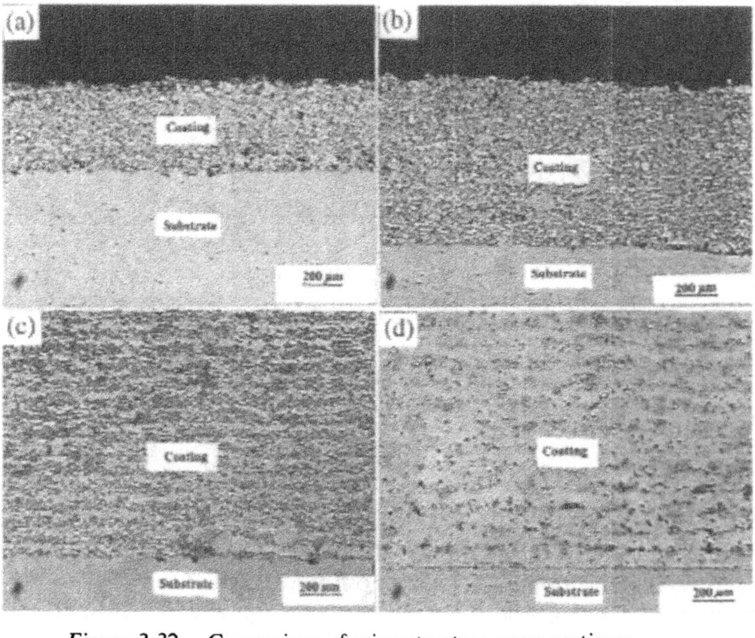

Figure 3.32. Comparison of microstructure cross-sections: (a)Mo-10% Cu; (b)Mo-20% Cu; (c)Mo-40% Cu; (d)Mo-70% Cu.

2.3.4 Friction Coefficient Measurement

A sliding tester using a ball-on-plate technique was used to assess the coefficient of friction. Figure 3.33 shows the relationship between the coefficient of friction and time. The figure shows that the 100% Mo test sample exhibits higher coefficients of friction compared with the rest of the test samples. The test sample with 90% copper added had the minimum friction values. Very stable values for the coefficient of friction have been observed for test samples produced with 40% copper. It has been noted that coefficient of friction values generally decrease with an increase in copper.

The test sample produced with 40% copper shows a stable coefficient of friction. Generally, it is thought that in plasma spraying temperature is not sufficient enough to melt the particles completely and this causes a high level of micro porosity along with inadequate adhesion to the substrate. Therefore, the laser irradiation operation during spraying increases the temperature, causing more melting of droplets during spraying. Self-annealing during spraying also enhances the splat orientation and this gives better contact and heat transfer with the oxide-free surface [55].

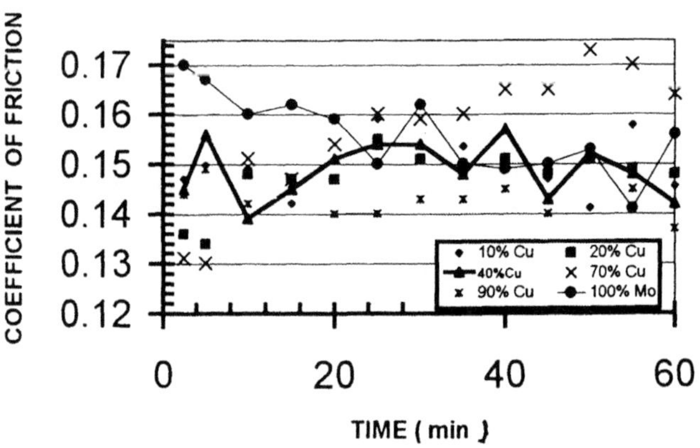

Figure 3.33. Comparison of coefficients of friction.

2.3.5 Evaluation of Wear Depths

The depths of the wear tracks were measured. It was observed that the sample with 70% Cu had the highest wear depth compared with the rest of the test samples. The minimum wear depth was observed for test samples produced with 40% copper, as shown in Figure 3.34.

The better stability could be attributed to the better morphology of the microstructure with a minimal level of micro-structural imperfections. The poor performance of the test sample with 70% copper could be attributed to the high level of micro-structural imperfections, as shown in Figure 3.32(d).

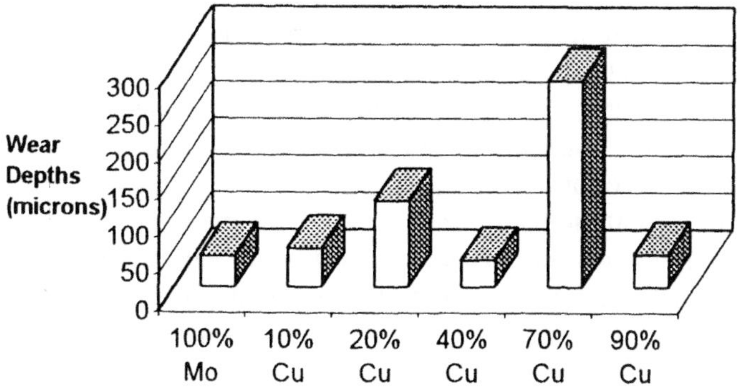

Figure 3.34. Comparison of wear depths.

2.3.6 Conclusions

Based on the results presented, the following conclusions can be drawn:

(1) Synthesized coatings have considerable potential as high-performance tribomaterials.

(2) The microstructural imperfections and the amorphous nature were shown to have an adverse effect on the tribological characteristics. The optimum properties were observed for a 40% copper addition.

(3) Laser irradiation during plasma spraying was shown to have a significant and favorable effect on microstructural aspects like microporosity, shrinkage voids, oxides and splats.

(4) The limitations associated with plasma coatings are significantly reduced by laser irradiation because the process involves inherently high temperatures and self-annealing occurs.

2.4 Martensitic and Reverse-Martensitic Transformation of Magnetically Graded Materials

2.4.1 Introduction

Many common austenitic stainless steels are mechanically metastable at low temperatures when subjected to sufficient stress or strain, and undergo spontaneous transformations into a ferromagnetic α' martensite. This martensitic transformation is accompanied by a paramagnetic to ferromagnetic transition [56,57]. Since the amount of the α' martensite increases with the increasing plastic strain, the magnitude of the saturation magnetization also increases with the increasing plastic strain. By making use of this phenomenon, a magnetically graded material could be fabricated by non-homogeneous deformation of the austenitic stainless steel [58-60]. This method is called the martensitic transformation technique.

On the other hand, it is also known that the ferromagnetic martensite phase transforms into the paramagnetic austenite phase when heated (reverse-martensitic transformation) [61]. In the range of the starting reverse-martensitic transformation temperature to the finishing temperature, the amount of saturation magnetization due to the martensite phase decreases with increasing temperature. Therefore, heat treatment in a thermal gradient will also change the saturation magnetization, in an amount of the martensite phase that depends on the local temperature (reverse-martensitic transformation technique) [62,63].

This section describes methods for fabricating magnetically graded materials using the martensitic transformation technique and the reverse-martensitic transformation technique. In addition, some advantages of the fabricated magnetically graded materials and the proposed fabrication methods are discussed.

2.4.2 Martensitic Transformation Technique

Figure 3.35 shows a sample of a magnetically graded material fabricated by non-homogeneous tensile deformation of austenitic stainless steel [58,59]. The distributions of the plastic strain and saturation magnetization are also shown in Figure 3.35. The saturation magnetization gradually increases with increasing distance over the entire length of the specimen. This is because of the deformation-induced paramagnetic γ to ferromagnetic α' transformation that occurs during the tensile deformation. With this method, one can successfully fabricate a magnetically graded material by the martensitic transformation technique.

In the same figure, the plastic strain as a function of distance has a positive curvature, even though the cross-sectional area of its reduced gauge section decreases linearly. Since the saturation magnetization versus plastic strain curve also has a positive curvature, the saturation magnetization as a function of distance along the specimen length must also have a positive curvature. In the tensile deformation of wedge-shaped specimens, it is difficult to predict the strain distribution along the specimen length [59]. Although the most important objective when fabricating an FGM is to achieve a well controlled distribution of functions (composition, texture, structure and other necessary elements), controlling the distribution of saturation magnetization by changing the local tensile strain is not considered to be very easy.

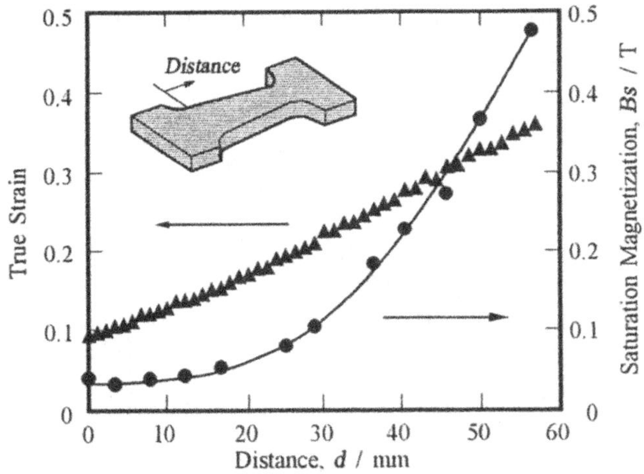

Figure 3.35. Example of wedge-shaped SUS304 stainless steel sample for non-homogeneous tensile deformation [58], and distributions of plastic strain and saturation magnetization within the sample.

When rolling a wedge shaped specimen, one can directly predict the strain distribution from the shape of the specimen. Rolling ratio distributions after the deformation in specimens were calculated and are shown in Figure 3.36. It should be noted that the curves have negative curvatures, even though the dimensions of the samples change linearly along the gradient. Since saturation magnetization as a function of the rolling ratio has a positive curvature, the saturation magnetization as a function of distance should have an intermediate curvature and will approach an approximately linear shape with a suitable selection of initial specimen dimensions. The saturation magnetization gradients in the specimens are also shown in Figure 3.36. It should be emphasized here that the relationship between the distance and the saturation magnetization is almost linear. Using this technique, a magnetic FGM with a linear magnetic gradient can be successfully fabricated by rolling wedge-shaped 304 stainless steel.

2.4.3 Reverse-Martensitic Transformation Technique

Figure 3.37 shows an example of a magnetically graded material fabricated using the reverse-martensitic transformation technique [63]. The variations of the local temperatures of the furnace during the thermal gradient heat treatment of the specimen are also shown in this figure. By substituting the relationship between the heating temperature and the saturation magnetization into the temperature gradient during the heat treatment, one can predict the distributions of the saturation magnetization in the specimen. It was found that the calculated saturation magnetization distribution in the specimen agreed with the measurements. Therefore, by

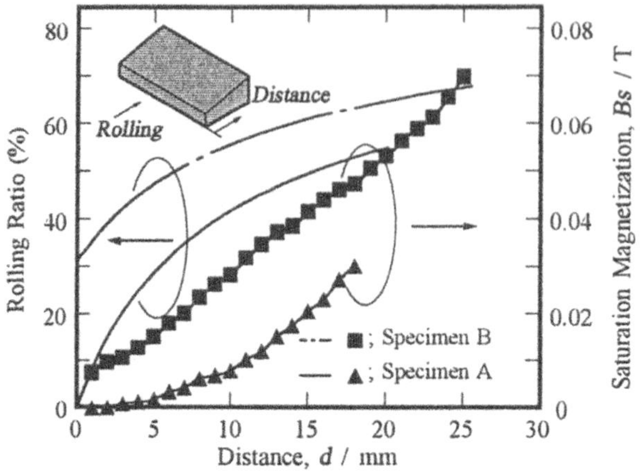

Figure 3.36. Rolling ratio distributions and saturation magnetization gradients after deformation in specimens [59], and shape of wedge-shaped sample.

controlling the thermal gradient in the furnace very carefully, one can fabricate a magnetically graded material with a suitable gradient of magnetization by using the α' to β reverse transformation with SUS304 stainless steel. It must be emphasized here that there are no limitations of initial and final shapes of the specimen for this technique. Therefore, for example, Tanaka *et al.* can even fabricate a magnetically graded fiber or wire using this technique [63].

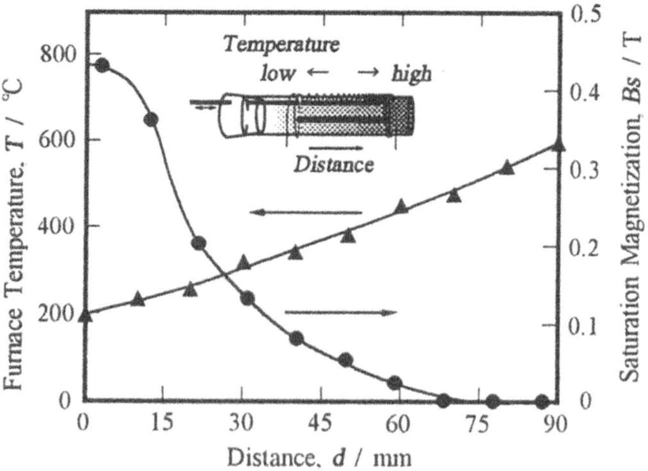

Figure 3.37. Variations of local temperatures in temperature gradient furnace (in which density of the heating coil was changed from place to place to introduce a temperature gradient), and saturation magnetization distribution in the heat-treated specimen [63].

2.4.4 Advantages of Magnetically Graded Materials

Current fabrication methods for FGMs are applications of a relatively new technology, and the fabrication facilities are expensive. In addition, it is difficult to produce relatively large materials with most of these methods. In contrast, no special facility is required for the processing techniques presented in this study. It is also easy to produce large items with these techniques.

A lot of effort is now being put into protecting the environment. New environment-friendly materials and products have also been developed. For this reason, ecomaterials are currently attracting a lot of attention. Ecomaterials are designed to minimize environmental impact while maximizing the material's performance and recyclability. It must be emphasized here that as magnetically graded material is a monolithic FGM, it is possible to return the material to its original state by heat treatment or hot working. Figure 3.38 shows the recycling process for a monolithic FGM produced using the martensitic transformation technique [64]. As can be seen, it is easy to reclaim virgin materials from FGMs, since deformation-induced martensite transforms into austenite when the deformed stainless steels are heated to the temperature of the austenite region. This may be one of the simplest methods for obtaining virgin materials from FGMs. This means that fabricated FGMs can be turned back to virgin materials using very easy methods such as melting or heat treatments. Therefore, one regards the fabricated FGMs as ecomaterials due to their easy and low-cost recyclability.

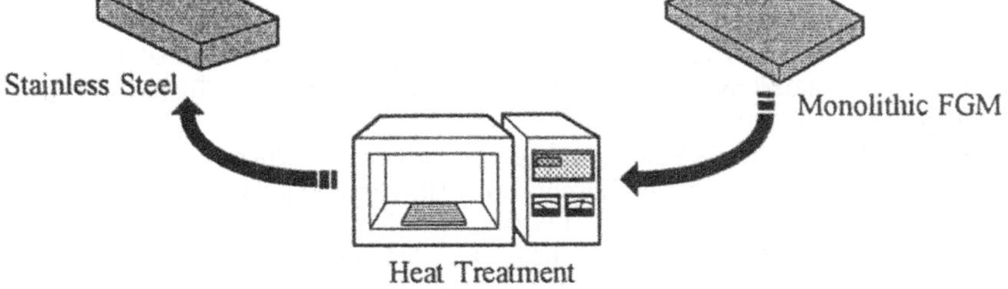

Figure 3.38. Recycling process for a monolithic FGM produced by the martensitic transformation technique.

Although the data is not presented here, the fabricated material has both magnetic and mechanical gradients. Moreover, it is known that the corrosion resistance of the austenitic phase is better than that of the martensitic phase. One can therefore simultaneously fabricate three different gradient properties in one sample. The magnetically graded material may be usable as a position-measuring device by combining it with a magnetic sensor. For instance, it can be used as a device that determines the focal point in an automatically focusing camera.

REFERENCES

1. Sata N., Characteristics of SiC-TiB_2 composites as the surface layer of SiC-TiB_2-Cu functionally gradient material produced by self-propagating high-temperature synthesis, Functionally Gradient Materials, edited by Holt, J.B., Koizumi, M., Hirai, T. and Munir, Z. A., Ceramic Transactions, 24 (1993) 109-116, The American Ceramic Society, Westerville, OH.
2. Munir, Z. A., Field effects in self-propagating solid state reactions. Z. Physik. Chem., 207 (1997), 39-57.
3. Feng, A., Munir, Z. A., Field-assisted self-propagating synthesis of beta SiC, J. Appl. Phys., 76 (1994), 1927-1928.
4. Feng, A. and Munir, Z. A., The effect of an electric field on self-sustaining combustion synthesis, Part I: modeling studies, Metall. Trans., 26B (1995), 581-586.
5. Xue, H. and Munir, Z. A., Synthesis of AlN-SiC composites and solid solutions by field-activated self-propagating combustion, J. Euro. Ceram. Soc., 17 (1997), 1787-1792.
6. Orru, R, Cao, G. and Munir, Z. A., Field activated combustion synthesis of titanium aluminides, 30A (1999), 1101-1108.
7. Gedevanishvili, S. and Munir, Z. A., The influence of an electric field on the mechanism of combustion synthesis of tungsten silicides. J. Mater. Res., 10 (1995),2642-2647.
8. Shon, I. J. and Munir, Z. A. Synthesis of TiC and TiC-Cu composites and TiC-Cu functionally graded materials by electrothermal combustion, J. Amer. Ceram. Soc., 81 (1998), 3243-3248.
9. Munir, Z. A., Shon, I. J. and Yamazaki, K., Simultaneous synthesis and densification by field activated combustion. US Patent, No. 5,794,113; August 11, (1998).
10. Carrillo-Heian, E.M., Ph D Thesis, University of California, Davis, 2000.
11. Halverson, D. C., Lum, B. and Munir, Z. A., The combustion synthesis of boride composites, Proceedings of High-Temperature Materials Chemistry Symposium IV, edited by Munir Z. A., Cubicciotti D, and Tagawa H. Proceedings, The Electochemical Society, Pennington, NJ, 88-5 (1988), 613-622.
12. Tokita,M., Trends in advanced SPS systems and FGM technology. Proceedings of the NEDO International Symposium on Functionally Graded Materials; Tokyo, Japan, October 21-22, (1999).
13. Readey, D. W., Colorado School of Mines, private communications.
14. Yamada, O., Miyamoto, Y. and Koizumi ,M., Self propagating high temperature synthesis of the SiC, J. Mater. Res., 1 (1986), 275-279.
15. Graeve, O. A., Carrillo-Heian, E. N., Feng, A. and Munir, Z. A., Modeling of wave configuration during electrically-ignited combustion synthesis. J. Mater. Res., submitted, (2000).
16. Ho, P. S. and Kwok, T., Electromigration in metals, Rep. Prog. Phys., 52 (1989), 301-348.
17. Bertolino, N., Garay, J. and Munir, Z. A, to be published.

18. Shon, I. J., Munir, Z. A., Yamazaki, K. and Shoda, K., Simultaneous synthesis and densification of $MoSi_2$ by field activated combustion. J. Amer. Ceram. Soc., 70 (1996), 1875-1880.
19. Ishiyama, M., Proc. 1993 Powder Metall. World Congress, Ed. by Y. Bando and K. Kosuge, Kyoto, Jpn. Soc. Powder & Powder Metall. Japan , (1993), 931-934.
20. Tokita, M., J. Soc. Powder Technology. Jpn., 30 (1993) 790-804.
21. Omori, M., Mater. Sci. & Eng. A, to be published.
22. Omori, M. Okubo, A. Gilhan, K and Hirai, T., J. Mater. Synthesis & Processing, 5 -4 (1997) 279-282.
23. Schreoter, K., U. S. Patent, No.1,549,615, (1925).
24. Suzuki, S., *Fabrication of Hard Metal*, Proc. 1st Sym. on Spark Plasma Sintering, Sendai, Japan , (1996) , 1.
25. Omori, M., Kakita, T., Okubo, A. and Hirai, T., J. Jpn. Inst. Met., 62 (1998) 986-991.
26. Antis, G. R., Chantikul, P., Lawn, B. R. and Marshal, D. B., J. Am. Ceram. Soc., 64-9 (1981) 533-538.
27. Kimura, H. and Kobayashi, S., J. Japan Soc. Powder & Powder Metall., 39-4 (1992) 287-290.
28. Omori, M., Sakai, H., Okubo, A., Kawahara, M., Tokita, M. and Hirai, T., J. Jpn. Soc. Powder & Powder Metall., 41 (1994) 649-652.
29. Omori, M., Sakai, H., Okubo, A., and Hirai, T., Proc. 3rd. Int. Symp. Structural & Functional Gradient Mater., edited by Ilschner, B. and Cherradi, N., Press Polytechniques Univ. Romandes, Switzerland, (1994), 667-671.
30. Omori, M., Okubo, A., Kan, G. H. and Hirai, T., Proc.4th. Int. Symp. Functionally Graded Materials, edited by Shiota, I. and Miyamoto, Y., Elsevier, Amsterdam, (1996), 767-772.
31. Tokita, M., 5th. Int. Symp. Functionally Graded Materials, edited by Kaysser, W. A., Trans Tech Pub., Germany, (1999), 83-88.
32. Omori, M., Kakita, T., Okubo, A. and Hirai, T., 5th. Int. Symp. Functionally Graded Materials, edited by Kaysser, W. A., Trans Tech Pub., Germany, (1998) , 53-58.
33. Gurland, J. and Bardzil, P., J. Met., 7 (1955) 311-315.
34. Tokita, M.., Trends in Advanced SPS Spark Plasma Sintering Systems and Technology, Journal of the Society of Powder Technology Japan, 30- 11 (1993), 790-804.
35. Omori, M., Sakai, H., Hirai, T., Kawahara, M. and Tokita, M., Preparation and Properties of ZrO_2(3Y)/Ni FGM., Proceedings of the 3rd International Symposium on Structural and Functionally Gradient Materials, Lausanne, Switzerland, (1994), 71-76.
36. Tokita, M., Development of Large-Size Ceramic/Metal Bulk FGM Fabricated by Spark Plasma Sintering, Proceedings of the 5th International Symposium on Functionally Graded Materials FGM 98, Dresden, Germany, (1999) , 83-88.
37. Brindley, W. J., Miller, R. A. and Aikin, B. J., Improved Bond-Coat Layers for Thermal-Barrier Coatings, *NASA Tech Briefs*, August (1998), 63-65.
38. Aboudi, J., Pindera, M-J. and Arnold, S M., Higher-Order Theory for Functionally Graded Materials, *Composites: Part B (Engineering)*, 30-8 (1999), 777-832.
39. Pindera, M-J., Aboudi, J. and Arnold, S. M., Thermomechanical Analysis of Functionally Graded Thermal Barrier Coatings with Different Microstructural Scales, *J. American Ceramics Society*, 81-6 (1998), 1525-36.
40. Pindera, M-J., Aboudi, J. and Arnold, S. M., The Effect of Interface Roughness and Oxide Film Thickness on the Inelastic Response of Thermal Barrier Coatings to Thermal Cycling, *Materials Science & Engineering A*, (2000), (in press). See also *NASA TM 1999-209770*, (1999).
41. Aboudi, J., Micromechanical Analysis of Thermo-Inelastic Multiphase Short-Fiber Composites, *Composites Engineering*, 5-7 (1995), 839-850.
42. Freborg, A.M., Ferguson, B. L., Brindley, W. J. and Petrus, G. J., Modeling Oxidation Induced Stresses in Thermal Barrier Coatings, *Materials Science & Engineering*, A245 (1998), 182-190.

43. Suresh, S., Giannakopoulos A. E., Alcala J.,Spherical indentation of compositionally graded materials: theory and experiments, *Acta Mater.*, 45 (1997), 1307-21.
44. Suresh, S. and Mortensen, A., *Fundamentals of Functionally Graded Materials*, IOC Communications Ltd., London, (1998).
45. Giannakopoulos, A.E. and Suresh, S., Indentation of solids with gradients in elastic properties: part II, axisymmetric indentors, *Int. J. Solids Structure* 33 (1997), 2393-28.
46. Sampath, S., Herman, H., Shimoda, N. and Saito, T., Thermal Spray Processing of FGMs, *MRS Bulletin*, January (1995), 27-31.
47. Sampath, S., Smith, W. C., Jewett, T. J., and Kim, H., Synthesis and characterization of grading profiles in plasma sprayed NiCrAlY-zirconia FGMs, *Mat. Sci. Forum*, 308-11 (1999), 383-88.
48. Kalman, R. E., A new approach to linear filtering and prediction problems, *Trans. of ASME—J. of Basic Engineering*, (1960), 35-45.
49. Aoki, S., Amaya, K., Sahashi, M. and Nakamura, T., Identification of Gurson's material constants by using Kalman filter, *Computational Mechanics*, 19 (1997), 501-06.
50. Gudmundsson, B. and Jacobson, B. E., Microstructure and Erosion Resistance of Vacuum Plasma Sprayed CoNiCrAlY/Al_2O_3 Composite Coating, Mat. Sci. Eng., vol. A108 (1989), 87-95.
51. Sampath, S. and Herman, H., Plasma Spray Forming Metals, Intermetallics and Ceramics, J. Met., 45-7 (1993), 42-49.
52. Sasaki, S. and Shimura, H., *et al.*, Proc. of ISTC 95, Kobe, Japan. (1995), 267-271.
53. Sampath, S. and Herman, H., Rapid Solidification and Microstructure Development During Plasma Spray Deposition, Jour. of Thermal Spray Technology, 5-4, Dec. (1996), 445-456.
54. Alam, S. and Sasaki, S, *et al.*, Intl., Conf., on Surface Modification technologies, Singapore, Sept. 2- 4, (1996), 918-924.
55. Sasaki, S. and Shimura ,H., *et al.*, Proc. 6th Intl., Trib. Cong. Aug 30, Budapest, Hungry.
56. Mangonon, Jr, P. L. and Thomas, G., *Metal Trans.*, 1 (1970), 1587-94.
57. Ding, J., Huang, H., McCormick, P.G. and Street, R., *J Mag. Mater.*, 139 (1995), 109-114.
58. Watanabe, Y., Nakamura ,Y., Fukui, Y. and Nakanishi, K., *J Mater. Sci. Letters*, 12 (1993); 326-28.
59. Watanabe, Y., Nakamura, Y. and Fukui, Y., *Proc. of FGM '96*, (1997), 713-718.
60. Watanabe, Y., Kang, S. H., Chan, J. W. and Morris, Jr, J.W., *Mater. Trans. JIM*, 961 40 (1999), 66.
61. Bourke, M., Maldonado, J. G., Masters, D., Meggers, K. and Priesmeyer, H. G., *Mater. Sci. Eng.*, A221 (1996), 1-10.
62. Sakai, H., Morishita, D. and Watanabe, Y., *Mater. Sci. Forum*, 308-311 (1999), 579-84.
63. Tanaka, F. and Watanabe,Y., *Proc. of Inter. Conf. on Adv. Fiber Mater.*, (1999), 391-392.
64. Watanabe,Y. and Fukui, Y., *Proc. of The Fourth Inter. Conf. on ECOMATERIALS*, (1999), 495-498.

Chapter 4

CHARACTERIZATION OF PROPERTIES

1. METHODS FOR EVALUATING FGMs

1.1 Introduction

Performance tests for FGMs include the local evaluation of the microstructure and material properties to manifest the performance of the designed structure and properties distribution. They also include the evaluation of the overall performance of FGM properties. In the microstructural evaluation, the quantitative determination of size, configuration, orientation and contiguity of phases is required. These may be possible by means of ordinary image analysis. Conventional apparatus may be applicable in some cases such as microscale chemical analysis and microhardness. However, in other cases some modifications might be necessary because of difficulties in measurement due to the continuous change of properties in a local region. On the other hand, in the evaluation of overall performance of FGM properties — such as electrical, magnetic, thermal and mechanical properties — ordinary methods may be applicable, however, some modifications or the development of new evaluation methods must be necessary. Moreover, fracture behavior in a ceramic/metal FGM may change from brittle to ductile fracture due to the gradual variation of the contiguity of the ductile phase, and metal phase. The overall mechanical behavior of the FGM has to be evaluated not only on macroscale but also on a microscale, such as change growth, micro-crack initiation at interfaces and crack propagation. Also, evaluation of thermal stress is important to analyze the mechanical response of the FGM components because they are generated during fabrications and heat treatments as well as in actual environments due to the large CTE differences of the elements. In many applications, FGMs may be exposed to constant or alternating thermal loading with a high temperature gradient from one side to the other. Thermal stability of the FGM structure, therefore, must be evaluated, because microstructural change may occur due to Ostwald ripening and because the compositional distribution may be unstable due to the diffusion of structural elements. Also, at the evaluation temperatures with rapid heating and cooling, thermal shock resistance, thermal fatigue characteristics, temperature profile and overall heat flow must be evaluated.

Although performance tests for FGMs have to be urgently developed, only a few methods are available at present which have been featured in some scientific reports addressing this topic. In the following sections some performance tests currently available will be briefly described in addition to a few nondestructive tests.

1.2 Elastic Property

There has been a successful evaluation of the elastic constants and its distribution of sintered FGMs of zirconia/stainless steel using an LFB acoustic microscope [1]. The results on the applicability of this method to uniform composites having specific sintered microstructures have revealed that the method enables the measurement of local elastic constants, such as Young's modulus and Poisson's ratio, of the sintered metal/ceramic composite with sufficient accuracy. It has been demonstrated that evaluation of the elastic constant distribution of FGMs, which have definite compositional gradients and pertinent microstructural transitions, was successfully achieved with the LFB acoustic microscope.

1.3 Thermo-Physical Property

The temperature response and overall thermal conductivity of FGM in pulse-wise heating was reported in an approximate solution which was defined from the analytical solution of the temperature response in the multilayer component during transient thermal loading [2]. The results have revealed that the approximate solution can be used to evaluate the heat-shielding properties if a maximum error of about 6% is acceptable. From an engineering point of view, the approximate solution is anticipated to be used not only for qualitative but also quantitative prediction of the temperature response of FGMs because of its simplicity and fair degree of agreement.

1.4 Thermal Stability

A thermal stability test was reported for plasma spray-coated FGM heat treatment under a constant and gradient temperature field [3]. The zirconia phase was resintered during the heat treatment of an uniform temperature field of 1473K. Vertical cracks formed in the FGMs due to the non-homogeneous shrinkage of the zirconia phase, in which Fe rich metal was infiltrated, much like a self-restoration. No change of the FGM structure was observed during heat treatment of a gradient temperature field of 873 to 1473K. Numerical evaluation of the thermal stability of the gradient structure based on the finite difference method has been applied to FGM exposed to steady-state and transient thermal loading [4]. The material combination of βTi/TiC and V/V_2C were used as the starting materials. The report has shown that condensations of carbide at the high temperature side for βTi/TiC and at the low temperature side for V/V_2C are caused by the difference of heat transport and partial molar enthalpy of solution. Growth of single phase was observed in the high and low temperature regions, which leads to breaking the gradient structure gradually.

1.5 Thermal Shock and Thermal Fatigue

Figure 4.1 shows a predictive testing method for FGM development focused particularly on the determination of the overall thermal barrier characteristics [5]. The first step includes fundamental testing for the determination of basic thermomechanical properties of homogeneous composite materials. Performance tests of thermal shock and thermal fatigue fall into two classes: small-scale basic evaluation tests and large-scale tests simulating industrial applications.

The laser heating method was applied to evaluate the thermal shock resistance of the ZrO_2 plasma-sprayed coating [5]. A CO_2 laser was used for irradiating the surface of the FGM specimen. The critical power density Pc at the first occurrence of vertical cracking, which is the critical laser power divided by the spot area, is determined to characterize the thermal shock resistance. A value three times higher than that of the non-GM was emphasized, indicating the effectiveness of FGM to enhance thermal shock resistance. A plasma-arc heating method was used for evaluating the oxidation/thermal shock resistance of a ZrO_2-based FGM prepared by a plasma-spray coating [6]. Ar plasma with the addition of a controlled quantity of oxygen was impinged on the FGM surface, while the bottom surface was cooled with Ar gas flow to induce the temperature gradient in the specimen. The oxidation rate of the coating system was characterized by a parabolic law independent of different graded compositions, and the

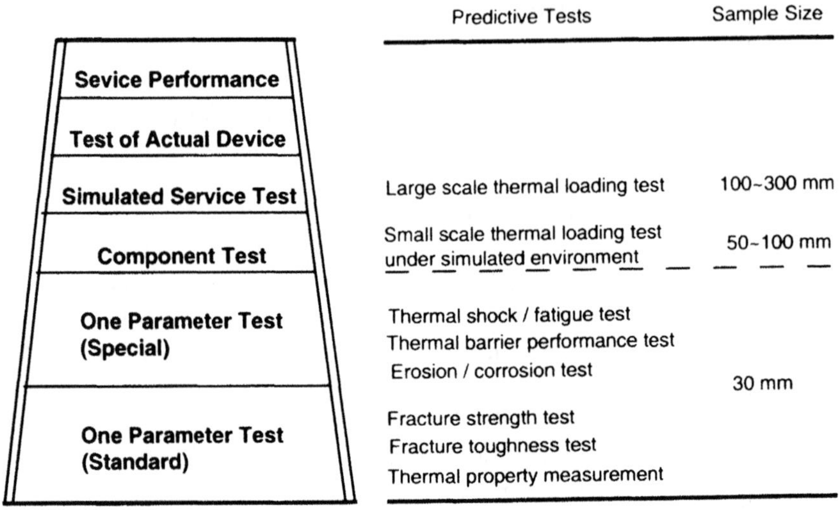

Figure 4.1. Predictive testing for FGM development.

oxidation behavior is chiefly controlled by the formation of alumina film in NiCrAlY.

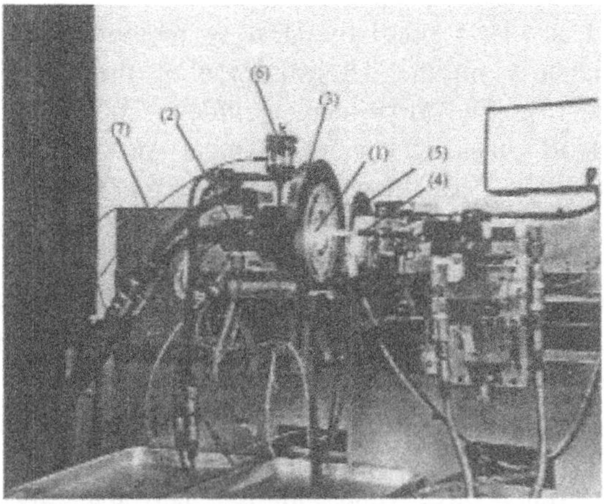

(1) TBC sample, (2) cooling chamber, (3) protect plate, (4) burner torch, (5) shutter, (6) sensor, (7) thermocouples.

Figure 4.2. Test system and sample setting configuration for burner heating test.

The results suggest that the use of the FGMs was effective for improving the thermal shock/fatigue resistance of the coating system. A burner heating method was applied to evaluate the thermal shock and fatigue (Figure 4.2) and the leading edge in an advanced gas turbine blade [7]. The FGMs were prepared by a powder metallurgical process and plasma spray-coating process. Acoustic emission technique can be combined with the testing method to detect the onset of cracking as well as damage such as delamination and spallation. The temperature for the first crack formation was defined as a thermal barrier performance value for the test sample. Note that the crack formation temperature was nearly constant for the carious sample and heating conditions indicating a material dependency of the thermal barrier property. The fracture mechanism during the burner heating test was discussed on the basis of thermal stress analysis by FEM. A fracture mechanism map was proposed for correlating the temperature condition with the coating damage, emphasizing the higher spalling resistance of ZrO_2-based FGMs over the conventional ZrO_2 single-layer coating.

Figure 4.3 shows variations of AE hits during heating and cooling as a function of thermal cycles for the corresponding specimens of type-1 [8]. The first increase of AE hits during cooling indicates mainly orthogonal crack formation and growth. After about 20 thermal cycles AE hits during heating appear to be detected which may be related to the formation of small transverse cracks in the graded layers. It should be noted that the final rapid increase of AE hits might be related to the formation of a large delamination crack by the linking up of the small transverse cracks. The variation of effective thermal conductivity is also plotted in Figure 4.3.

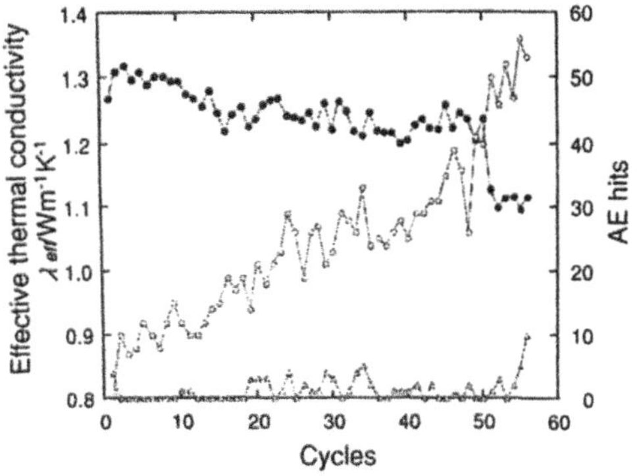

Figure 4.3. Variation of AE hits during heating and cooling as a function of thermal cycles the corresponding specimens of type-1.

The effective thermal conductivity of a specimen is defined as,

$$\lambda_{eff} = q/[(T_s - T_b)/t] \tag{4.1}$$

where q is the heat flux loaded to the specimen. Ts is the averaged top surface temperature and Tb is the bottom surface temperature of the specimen. The effective thermal conductivity decreases moderately with the increase in the number of thermal cycles and suddenly drops largely about 50 thermal cycles. It is known that the change of effective thermal conductivity is useful in examining the thermal fatigue characteristics of a coating, because cracking in a coating generally causes the effective thermal conductivity to decrease. The transverse crack, since it is oriented in the normal direction to the heat flow, more effectively decreases the effective thermal conductivity. Thus, the sudden drop of the effective thermal conductivity may well indicate the formation of a delamination crack. Consequently, the spallation life of the coating can be defined as the critical number of the thermal cycle, evaluated from either the rapid increase of AE hits during heating or the sudden drop of the effective thermal conductivity.

The number of cycles to failure by a given temperature difference pertinent to an applied heat flux, is used to compare the spallation life of the TBCs under thermal cycling. The tests were completed when a large increase of AE hits during heating was detected. Figure 4.4 shows the relationship between the temperature difference in a coating and the logarithmic number of cycles to failure of the TBCs. It can be noted that the failure life increases linearly with the decrease of the applied thermal load.

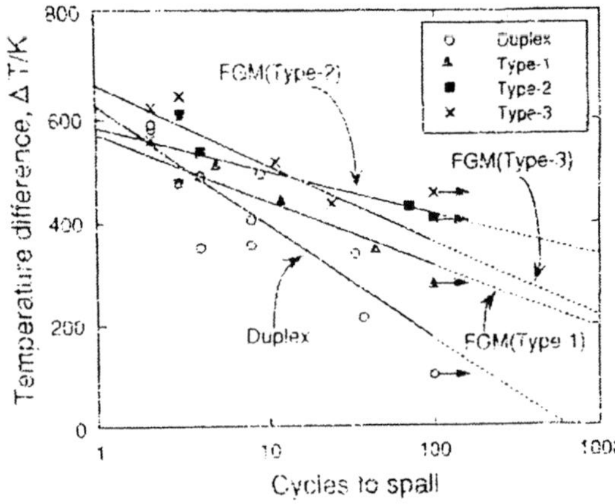

Figure 4.4. Spallation life of the TBCs under thermal cycling.

This relationship is similar to that of the fatigue behavior of a metallic material: that is, the fatigue life is dependent on the mean applied stress, which is well known as an S-N curve. It is clear from this figure that the FGM coatings show much longer failure life than conventional duplex coatings. For example, at a given temperature difference of 300 K, FGM type-1 survives more than 200 cycles. Compared with the duplex TBCs, the results have shown that functionally graded TBCs possess the desirable effect of improving failure life under cyclic thermal loads. On the other hand, it is important to note that the failure life of functionally graded TBCs depends on the compositional profile of the gradient. Among the functionally graded TBCs, FGM type-2 may have good potential to give a longer failure life. At a given temperature difference of 400 K, FGM type-2 is expected to survive ten times more cycles than FGM type-1. Therefore, the FGM type-2 with a linear compositional profile might be the best one to improve failure life under thermal cycling.

An electron beam heating test was applied to evaluate the W/Cu gradient material prepared by the infiltration process as a beam target against plasma [9]. The pre-designed operating condition of 15 MW/m^2 was achieved without any damage while radiating the electron beam on the top surface of the specimen.

The large-scale tests simulating actual applications include Xe-arc lamp heating test, high-temperature gas heating tests for FGM applications in scram jet engines, aerodynamic heating field test under such condition as re-entry into the atmosphere, and high-speed rotational heating test for FGM application in supersonic turbo jet engine parts. These tests [10] were devised mainly for evaluating thermal barrier properties in a space plane project in Japan and the research is still under way.

1.6 Nondestructive Evaluation

For the practical application of FGM as functional components and structural materials, quality control of FGMs is indispensable. Non-destructive evaluation of the structural integrity of FGMs is required during fabrication, heat-treatment, assembly and in actual environments. Non-destructive evaluation is generally classified into two groups: material evaluation, which involves characterization of materials, degradation and properties, and flaw detection such as micro-cracking and delamination. Up to the present, a variety of non-destructive test methods have been proposed such as X-ray, ultrasonic, acoustic emission and eddy current, however, few methods have been applied to FGMs (see Refs.[2], [11], [12], [13]).

1.7 Summary

Evaluation methods for functionally graded materials have been briefly described. Although performance testing for FGMs has to be urgently developed, only a few methods are available at present, in particular for thermal shock resistance, elastic constant distribution and thermo-physical properties.

Functionally graded materials are developed through an iterative feedback loop involving material design, fabrication and evaluation. However, the final validation of fabricated FGMs is manifested through performance tests. Standardized performance tests for FGMs must be well established as soon as possible and necessitates research based on international collaboration.

2. PHYSICAL PROPERTIES

2.1 Ultrasonic Imaging

2.1.1 Introduction

Since the mechanical and physical properties of functionally graded materials (FGMs) are very susceptible to their internal microstructures, the establishment of a nondestructive method to evaluate non-homogeneous texture has become an important subject for assessing both the reliability and quality of FGMs.

Takatsubo *et al.* have recently developed a digital ultrasonic imaging system for visualizing and analyzing ultrasonic waves propagating on solid media. Using this imaging system, ultrasonic waves propagating on a metallic-ceramic FGM are observed, and the applicability to nondestructive characterization of the FGM is examined. This imaging technique can be an effective means for characterizing the internal microstructures in FGMs.

2.1.2 Laser Ultrasonic Method

2.1.2.1 Optical Imaging System

An ultrasonic imaging system developed newly by Takatsubo *et al.* is shown in Figure 4.5. An ultrasonic probe was mounted on a FGM as shown in Figure 4.5. The nominal frequency of the probe used in the measurement was 10 MHz, and its outside diameter was 10 mm. Ultrasonic short pulses were emitted from the ultrasonic probe repeatedly, and the displacements of the surface waves were detected by scanning a laser optical probe with a pitch of 0.1 mm. The detected displacement data were stored on a computer's hard disk. After the measurement, the stored data were reconstructed as animation images of the surface wave propagations. Using this imaging system, they can measure nano-meter order displacements in a frequency range up to 50 MHz with a minimum pitch 10 μm.

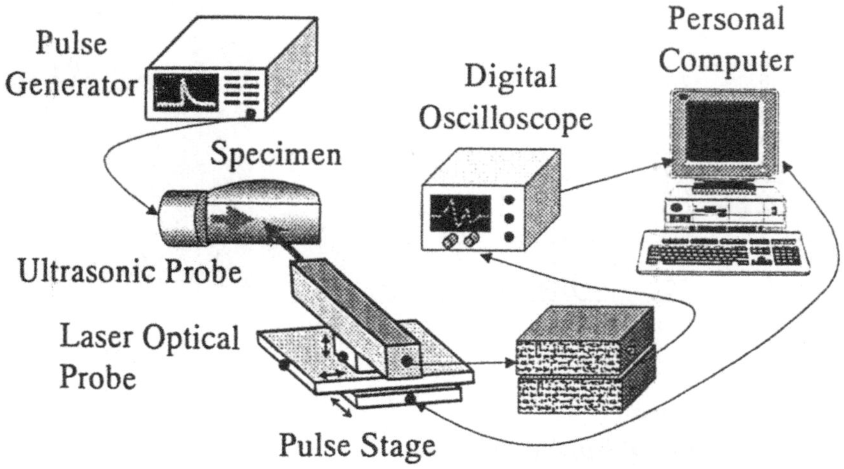

Figure 4.5. Optical heterodyne interferometer system for visualization of ultrasonic waves.

2.1.2.2 *Specimens*

The specimens were sintered from laminated powders of zirconia (ZrO_2) and stainless steel (SUS410) by the spark plasma sintering method. Figure 4.6 shows the dimensions of the specimens. The powder compositions of three samples are presented in Table 4.1.

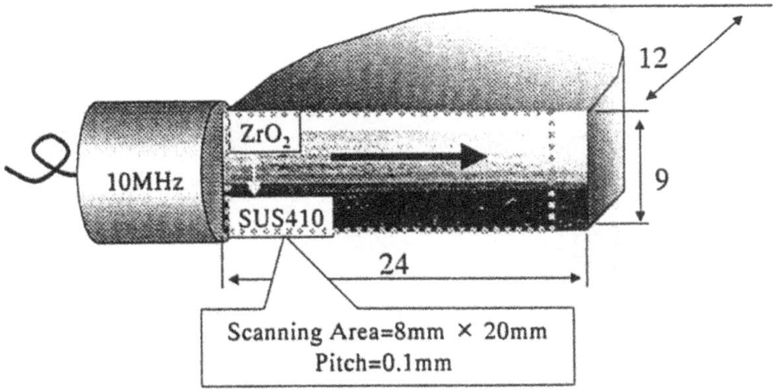

Figure 4.6. Dimension of ZrO_2/SUS410 FGM and ultrasonic position.

Table 4.1. Powder compositions of ZrO_2/SUS410 FGM.

SUS/ZrO2 (Vol%)	Weight(g) No.1	No.2	No.3	Thickness (mm)
100/0	16.49	16.49	16.49	3
70/30	2.57	2.57	2.57	0.5
60/40	2.51	2.51	2.51	0.5
50/50	2.45	2.45	2.45	0.5
30/70	2.33	2.33	2.33	0.5
20/80	2.27	2.27	2.27	0.5
10/90	2.21	2.21	2.21	0.5
6/94	0	0	2.19	0.5
5/95	0	2.19	2.19	0.5
3/97	0	0	2.18	0.5
0/100	6.44	6.44	6.44	1.5

2.1.2.3 *Images of Surface Waves*

Figure 4.7 shows a typical example of the visualization of surface wave which was propagating on specimen No.1. Figure 4.8 shows the B-mode images at lines A and B in Figure 4.7. The slopes of the inclined lines in Figure 4.7 represent the velocities of the surface waves. Thus, the velocity of the surface waves propagating on the FGM can be easily measured. As can be seen in Figure 4.6, the sound velocities at the middle depth of the specimen are slow compared to those at both sides of the monolithic zirconia and stainless area. Takatsubo *et al.* found that this velocity reduction was caused by the existence of micro pores. Actually, a very large number of pores were observed at the middle depth of the FGM as shown in Figure 4.9.

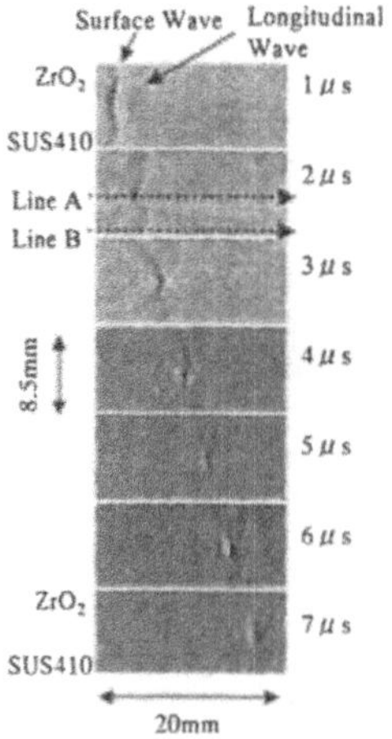

Figure 4.7. Visualization of surface waves propagating on ZrO_2/SUS410 FGM.

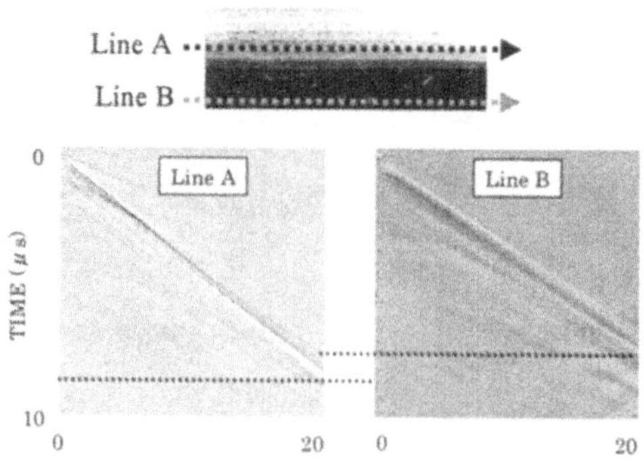

Figure 4.8. B-mode images at line A and line B.

The porosity distribution was measured with a microscope image analysis and is shown in Figure 4.10. As has been already reported [14, 15], the relationship between porosity, ψ, and ultrasonic velocity, V, is expressed as

$$\psi=4[1/\{(\phi_{zr}/V_{zr}+\phi_{sus}/V_{sus})\cdot V\}-1]/[3\{\pi\beta/(2\alpha)-1\}] \quad (4.2)$$

where, ϕ_{zr} and ϕ_{sus} are volume fractions of both ZrO_2 and SUS410, respectively. V represents the sound velocity. α is the reduction ratio of creeping wave speed [14]. β is the pore shape factor [14]. Hence, one can estimate the porosities from Equation 4.2, using the values of V_{zr} = 3220m/s, V_{sus} = 3050m/s, α=0.8 and β=2. The calculated values shown in Figure 4.10 compared with the measured values, which means one can estimate the porosity distribution in an FGM by the ultrasonic visualization method.

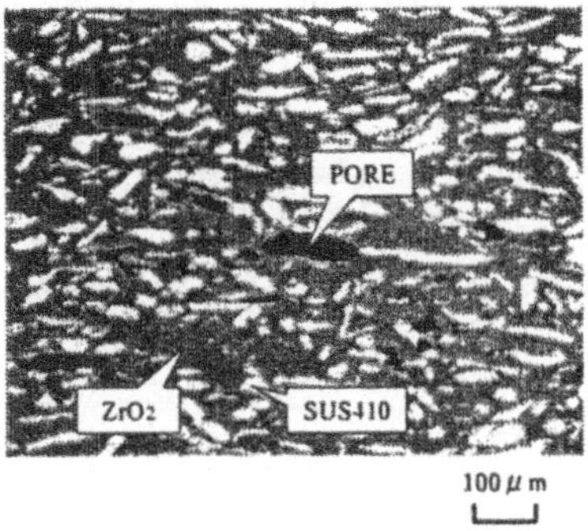

Figure 4.9. Pores observed in No.1 FGM.

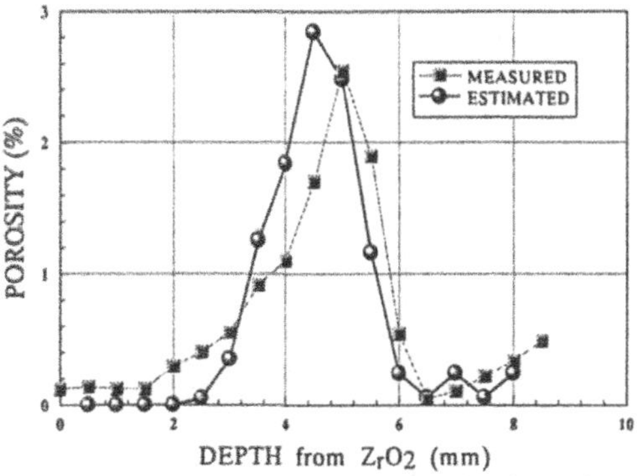

Figure 4.10. Measured and estimated porosities.

2.1.3 Leaky Rayleigh Wave Method

The leaky Rayleigh wave method is another effective means of measuring surface wave velocities. The basic concept [16] of measuring the leaky Rayleigh velocity is shown in Figure 4.11. By measuring the time difference Δt between the reflection wave and leaky Rayleigh wave, the leaky Rayleigh velocity, Vr, can be estimated from Equation 4.3,

$$Vr = Vw/\sqrt{(Vw\cdot\Delta t/z)-(Vw\cdot\Delta t/z)^2/4} \quad (4.3)$$

where Vw is the velocity of water. By scanning a cylindrical transducer, the velocities of the surface wave were measured.

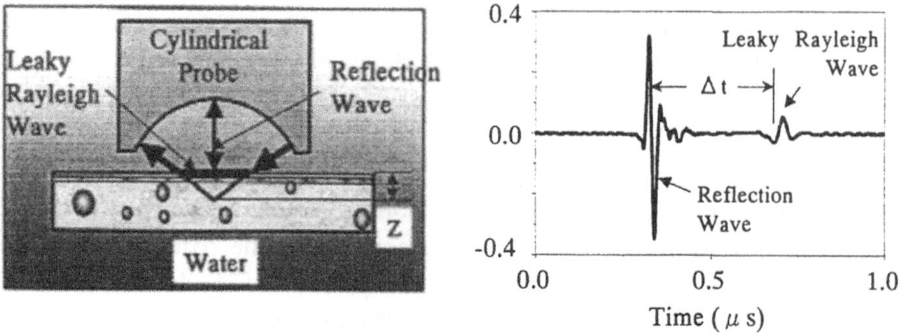

Figure 4.11. Velocity distribution of leaky Reyleigh waves.

Figure 4.12 shows the surface wave velocity distribution of specimen No. 3. From this figure, one can obtain the porosity distribution, as shown in Figure 4.13.

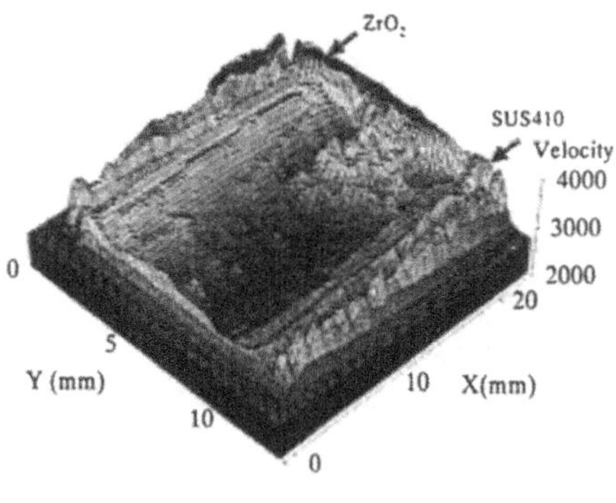

Figure 4.12. Velocity distribution of leaky Rayleigh wave.

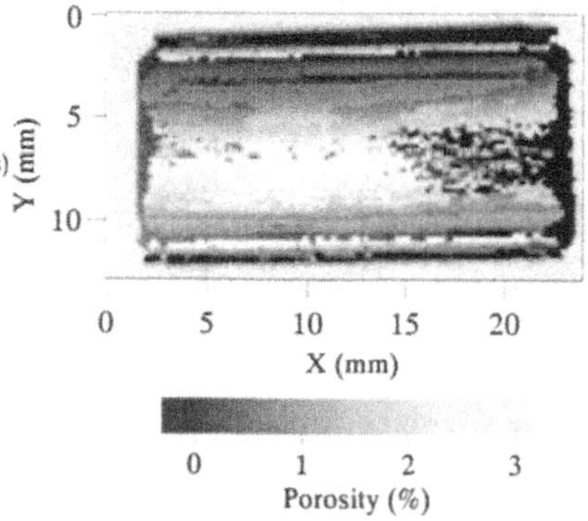

Figure 4.13. Porosity distribution of No.3 specimen.

2.1.4 Pulse Echo Imaging Method

A fully digital pulse echo imaging system in which all ultrasonic signals are digitized and passed to a computer hard disk have been developed.

As a typical pulse echo signal, a signal detected at the center position of specimen No. 1 is shown in Figure 4.14. The waveform has some reflecting echoes between the surface and bottom echoes, which correspond to laminate interface echoes. Figure 4.15 shows the B-mode images which display the vertical cross sections. In this figure, one can clearly observe the interface echo lines between the 100% SUS and 70% SUS area, which implies that the bonding qualities of those interfaces are weak compared with other interfaces. One can also obtain the horizontal cross section images, as shown in Figure 4.16. The black areas have high-amplitude echoes. These results imply that one can estimate the interface quality of metallic-ceramic FGMs.

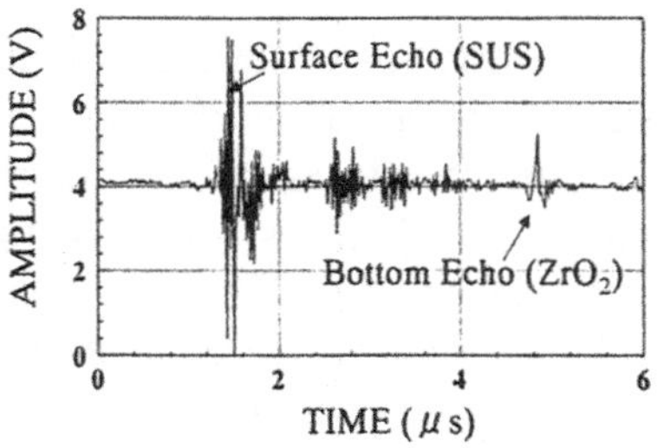

Figure 4.14. A typical pulse echo signal.

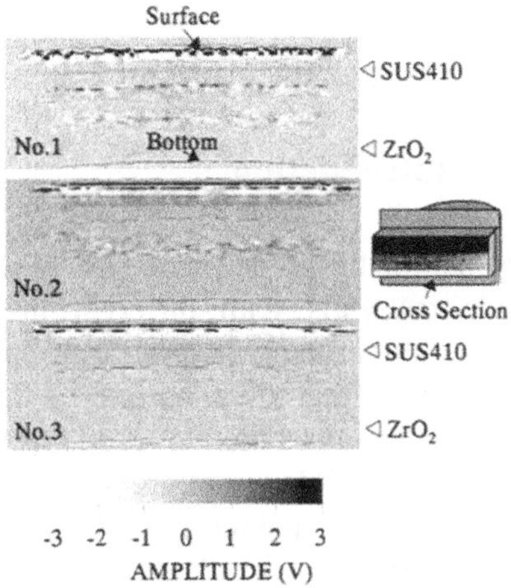

Figure 4.15. B-mode images (vertical cross section) of ZrO_2/SUS410 FGM.

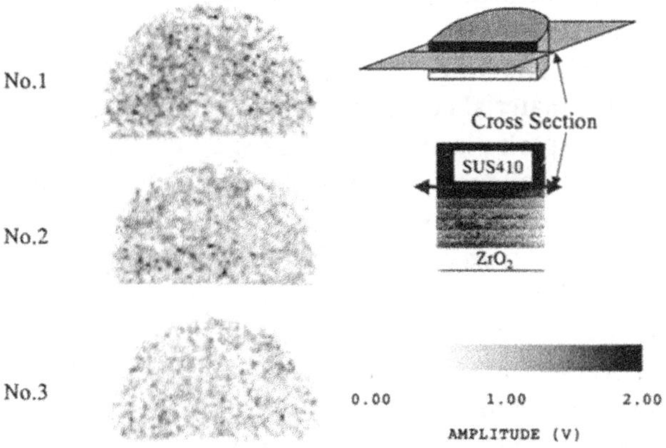

Figure 4.16. C-mode images (horizontal cross section) at the first echo.

2.1.5 Conclusions

The applicability of the ultrasonic full digital imaging method to nondestructive characterization of FGMs was discussed. The results are as follows:

(1) Ultrasonic waves propagating on a ZrO_2/SUS410 FGM can be visualized with the laser optical imaging system.

(2) The porosity distribution of the FGM can be estimated by the laser

ultrasonic method or the leaky Rayleigh method.

(3) The interface quality can be visualized by a digital pulse echo imaging method.

2.2 X-ray Diffraction

2.2.1 Introduction

Functionally graded materials (FGMs) have composite structures which include materials with different properties, therefore they usually have stresses inside their bodies, especially on the interfaces of the layers. Estimation of this stress is one of the important techniques for predicting their mechanical or thermal properties as a thermal barrier. The mechanical properties of FGMs are of great concern [17] because these materials possess the potential for a wide range of industrial applications [18]. Such multilayer materials are available for synthesis employing various coating and sintering techniques. One of the techniques is the plasma spray deposition method [19, 20], and CVD [21] or PVD [22] coating methods. On the other hand, sintering techniques for bulk materials including the HIP-sintering technique [23], hot-pressing technique[24], and self-propagating high-temperature synthesis [25] are also quite useful but they require long sintering time, special sintering techniques for preparation of powders, and limitations on the material combinations for sintering.

Recently, the spark plasma sintering (SPS) technique [26] has improved from several points of view and provides easy handling, and it is therefore expected to be used for industrial-scale applications because it has fewer limitations compared with other coating and sintering techniques. Using SPS, metal-ceramic bulk materials are easy to produce as FGMs. But SPS has not yet been established for industrial applications because the sintered materials have several types of stress in their bodies. These stresses are expected to strengthen their mechanical properties. For this reason, the stress in the SPS materials is an important factor to be evaluated.

2.2.2 Preparation and Measurement

A study was carried out to evaluate the stress in spark plasma sintered ZrO_2/SUS410L FGMs by employing the X-ray diffraction technique as lattice stresses and residual stresses. For the mechanical property of these materials, hardness was measured for the same bulk plane. The ZrO_2/SUS410L FGMs were provided by Izumi Technology Co., Ltd., and were sintered using the SPS technique. The ZrO_2/SUS410L FGMs had eight layers consisting of SUS410L as the substrate and ZrO_2 as the topcoat. These two layers sandwiched their mixtures which consisted of the weight ratio of 90 wt% to 50 wt% and 30 wt% of ZrO_2. The stresses and hardness were measured using 14 points of these FGMs. Basically the points were selected near the interfaces of layers and the center of layers. The mixture amounts were changed to control the graded functions of n = 0.5, 1.0, and 2.0. The total thickness was controlled by the SUS410L amount.

The stresses were measured with an X-ray micro-diffractometer (Rigaku Co., Ltd.; RINT 2000 system) using the Cr target ($K\alpha1$ = 0.22897 nm) at the power of 40 kV-300 mA. The X-ray was irradiated in the parallel direction of the layers and collimated to 30 μm in diameter. The lattice stresses were calculated from the lattice parameters of ZrO_2 and α-Fe. The lattice parameters were measured by an internal standard method using silicon powder at an incidence angle fixed on 30 degree in θ. The lattice parameters of the ZrO_2 were calculated by a least square method from 9 peaks. The lattice parameters of the α-Fe were calculated by an extrapolation method from two peaks of α-Fe. These lattice parameters were measured for all the graded functions. The residual stresses were measured for the graded function of 0.5 by an iso-inclination method. The residual stresses of the ZrO_2 were evaluated from two peaks. One is (113) peak (2θ= 101.564 deg.) at the incidence angles of 0 to 35 on every 5th degree. And the other was (004) peak (2θ= 128.704 deg.) at the incidence angles of 0 to 25 on every 5th degree. For the calculation of the residual stresses of ZrO_2, Young's modulus was assumed at 300 GPa, Poisson's ratio was assumed at 0.29, and stress constants were assumed at −35.69 MPa/deg. for (113) and −197.49 MPa/deg. for (004), respectively. The residual stresses of α-Fe were evaluated from (200) peak (2θ=106.029 deg.) at the incidence angles of 0 to 35 on every 5th degree. For the calculation of the residence stresses of α-Fe, Young's modulus was assumed at 300 GPa, Poisson's ratio was assumed at 0.28 and stress constants were assumed at −110.40 MPa/deg. The hardness was measured with a Vickers hardness tester (Akashi Co. Ltd., MVK-E) at 100 gf of loading weight.

2.2.3 Residual Stress of ZrO_2/SUS FGM

Optical microscope observations of cross sections of the three types of FGMs indicated that there were no cracks or voids inside the sintered bodies at a low magnification. From the X-ray micro-diffraction patterns of these FGMs, there were no phases except ZrO_2 and α-Fe phases. These results suggested that the SPS processing of these FGMs produced no new phases. One of the results of the lattice parameters is shown in Figure 4.17 in the case of the graded function of 0.5 as the shift ratio from the standard value in the JCPDS card. In this figure, the horizontal axis shows the position of the cross section as a zero position at the interface of the pure ZrO_2 and 90 wt% mixture. The *a*-axis lengths of the ZrO_2 and α-Fe were longer than those of the standard lengths, respectively.

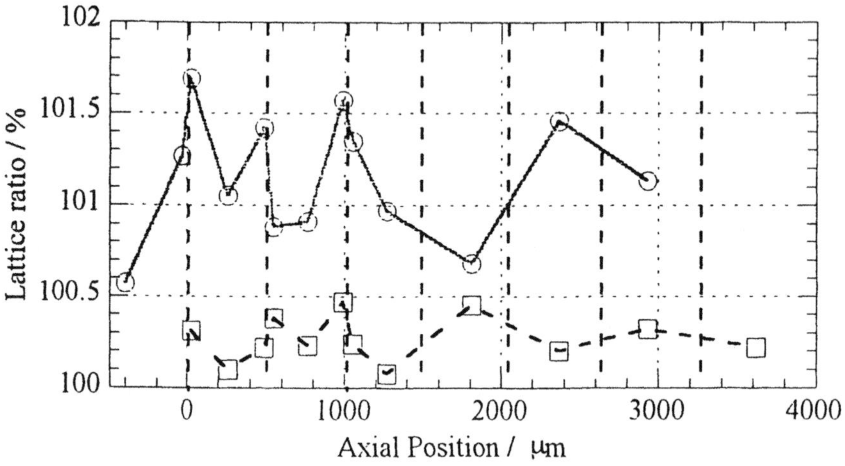

Figure 4.17. Normalized *a*-axis length of ZrO_2 (circle) and α-Fe (square) for n=0.5. Dotted lines show the interface position.

The normalized lattice parameters of all samples were independent with regards their compositions and graded functions. The lattice parameters, especially axis lengths, jumped up or down between the interfaces of the layers (dotted lines show the interfaces of the layers in Figure 4.17). These results might suggest that the lattice stresses existed in the ZrO_2 phases for all layers but the stresses were independent of the mixed ratio or axis direction. On the other hand, the lattice parameters of α-Fe were also longer than the length of the standard for all layers. These results might indicate that the lattice stresses of α-Fe were recognized in all layers, especially near the interfaces of layers, but the stresses were independent of the mixed ratio. These tendencies were observed in the other graded functions. These results might suggest that the lattice stresses were relaxed due to the increase of layer thickness. Figure 4.18 shows the residual stresses from (113) and (004) of ZrO_2, and (200) of α-Fe for each layer at the graded function of 0.5 (dotted lines show the interfaces of the layers in Figure 4.18). The residual stresses increased near the interfaces, and the direction of the stress changed near the interfaces. These results indicated that the interface of the layers kept the residual stresses. The results of the lattice stresses and residual stresses show that significantly higher values of stress had not been found in the cross sections of the test samples. It was also observed that at the ZrO_2 rich layers, the stress values increased near the interfaces and also the directions of the stress changed near the interface of layers. These results strongly suggest that the random stresses remained inside the sintered bodies, especially near the interfaces, without relaxing by the spark plasma sintering.

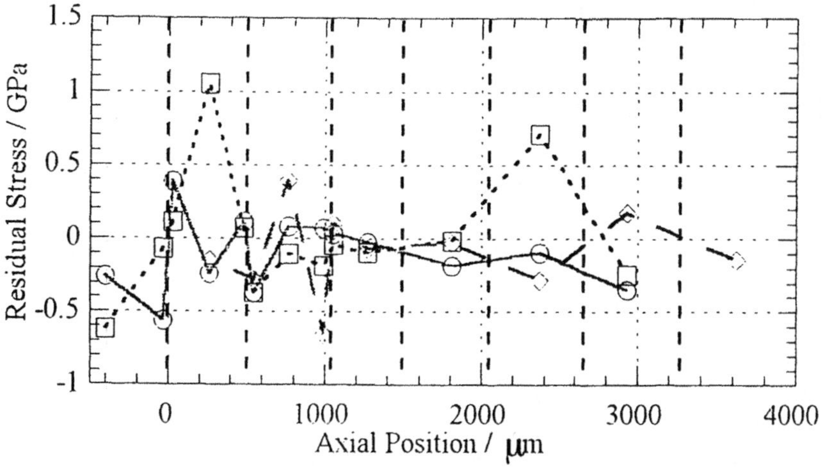

Figure 4.18. Residual stresses of ZrO_2 (113) (circle) and (004) (square), and α-Fe (200) (diamond). Dotted lines show the interface position.

The result of the Vickers hardness of the FGM at the graded function of 0.5 is shown in Figure 4.19. The hardness also changed between the interfaces.

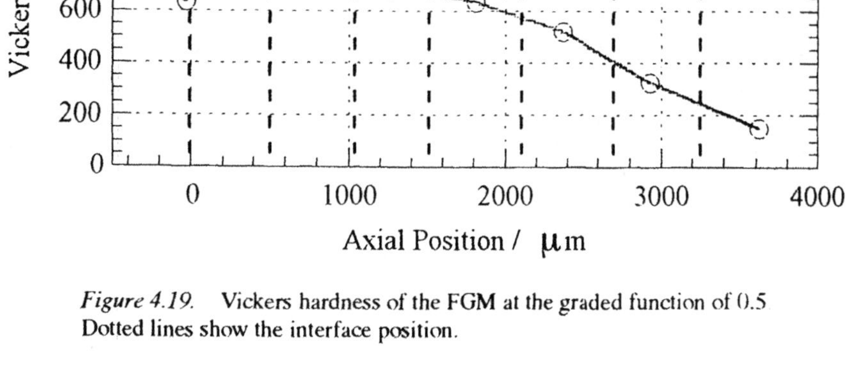

Figure 4.19. Vickers hardness of the FGM at the graded function of 0.5 Dotted lines show the interface position.

The hardness was reduced by increasing the SUS amount, except for the pure ZrO_2 layer, on the contrast of stresses. The main reason was thought to be that the hardness reflected the average properties of the crystal phases and several grains of the measurement area. The stresses from the XRD were measured to focus on the special phase or peak. These differences appeared in the results.

2.2.4 Conclusions

A study was carried out to evaluate the stresses in spark plasma sintered ZrO_2/SUS410L FGMs by employing the X-ray diffraction technique as lattice stresses and residual stresses. Hardness was also measured for the same bulk plane.

The residual and lattice stresses of the FGMs were evaluated from the lattice parameters of ZrO_2 and α-Fe for each layer. The residual stresses showed a tendency of changing their directions, tensile or compressive stress, at the interface between the layers. This result also strongly suggested that the interface of layers may be the cause of the residual stress and the trigger of a stress type conversion. The lattice parameters shifted from the standard values of both phases. The axis length of both phases changed in a wide range depending on the mixture ratio and jumped at around the interface between the layers. These results indicate that the lattice stress existed near the interface of layers more than near the center of the layers.

The stress was evaluated with regard to mechanical hardness. The Vickers hardness decreased with the ratio decrease of the ZrO_2. This macroscopic property did not appear only in microscopic stress but also in the interface contact force of ZrO_2 and α-Fe.

2.3 Micro-Raman

2.3.1 Introduction

There are many methods for evaluating the residual stresses in materials, among which the micro-Raman technique has the highest spatial resolution. However this technique is limited to materials that exhibit strong Raman scatter such as silicon. Zirconia, which is the base material of the FGM Mizuhara *et al.* interested in, also shows strong Raman scatter. Thus this technique to evaluate the residual stress in the FGM was tried.

2.3.2 Micro-Raman Technique

The micro-Raman technique is widely used for evaluating the residual stresses in silicon disks and Mizuhara *et al.* were very successful using it to evaluate the material. The technique has a spatial resolution of a few microns which is the best among the stress-measurement techniques. However the Raman scatter technique has several limitations, such as very low scatter efficiency or inapplicability to metals.

Fortunately, Zirconia (ZrO_2) the major material of the FGM to be interested in shows very strong Raman scatter and it has been reported that some of the peaks shift by stress [27]. They tried the developed method that measures the stokes and anti-stokes lines simultaneously [28]. Figure 4.20 shows the configuration of the Raman system.

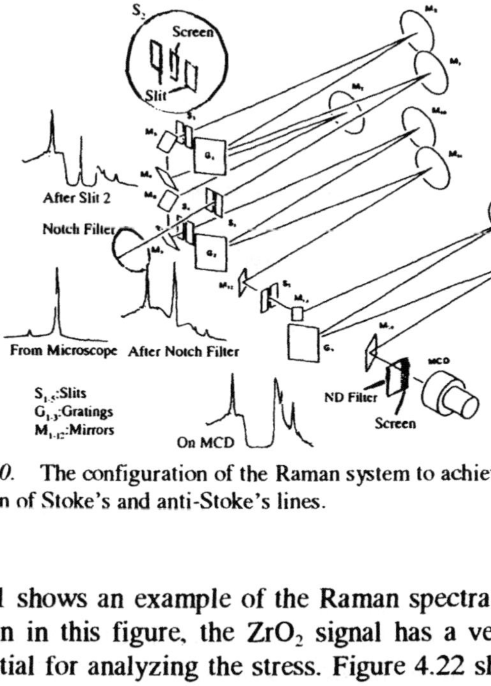

Figure 4.20. The configuration of the Raman system to achieve the simultaneous observation of Stoke's and anti-Stoke's lines.

Figure 4.21 shows an example of the Raman spectra obtained for ZrO_2. As can be seen in this figure, the ZrO_2 signal has a very good S/N ratio, which is essential for analyzing the stress. Figure 4.22 shows the four-point bending machine used to apply the loads. The compositions and the structure of the FGMs are shown in Table 4.2.

According to Ref. [27], zirconia's monoclinic peak at 614 cm^{-1} shifts 2.1 cm^{-1}/GPa, and tetragonal peak at 634 tetragonal cm^{-1} shifts 2.4 cm^{-1}/GPa respectively. Those shifts are smaller than that of silicon (4.4 cm^{-1}/GPa). However, one should be able to observe adequate peak shifts. Mizuhara *et al.* have measured the peak position of those peaks under 0, 166 and 333 MPa of stresses using the 4-point bending tester. The peak shift between 166 and 333MPa was 0.44 and 0.46 cm^{-1} and those are 1.2 times more than Ref.[27], but still reasonable. However the peak position of the unloaded specimen

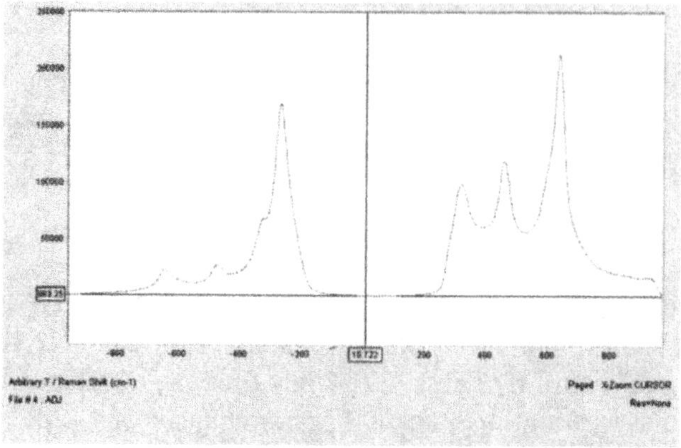

Figure 4.21. An example of Raman spectra obtained on ZrO_2.

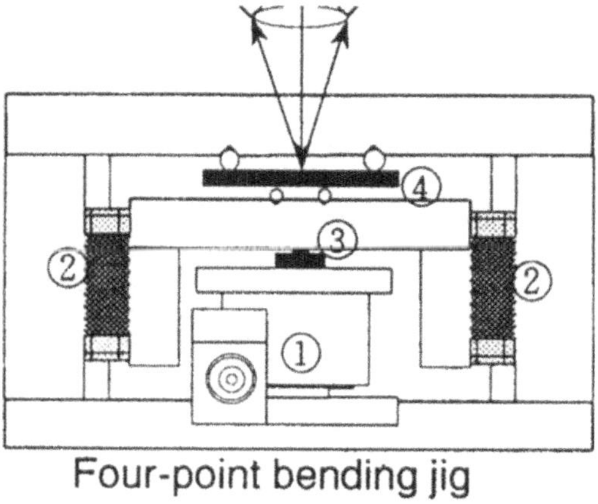

Figure 4.22. Four point bending tester.
1:Loader, 2:linear guide, 3:local cell, and 4:test piece.

scatters at around 0.2 cm^{-1}, and that corresponds to 100 MPa of stress, which is about three times worse than that of silicon. The reason for the large error is that the peak width of the ZrO_2 is broader than that of silicon and this significantly reduced the repeatability of the peak fittings. Also the relative peak intensity of the anti-Stokes peak of ZrO_2 is much smaller than that of silicon, thus the accuracy of the anti-Stokes peak position is worse. To avoid this problem they add an additional ND filter at the Stokes side and elongated the data sampling times. Figure 4.23 shows an example of residual stress distribution in FGM2. However, no reasonable data was obtained from the FGM1 in Table 4.2. The reason why the peak shift was not visible may be that there were lower residual stresses in FGM1 because of its lower density.

Table 4.2. Compositions of FGMs.

	FGM1	FGM2	FGM3
Sintering Temp. ℃	1130	1020	1020
Pressure MPa	50	25	25
Layers	11	11	8
ZrO_2%/Thick. mm	100/1	100/3	100/3
Layer 2	90/1	90/1	90/1.5
Layer 3	80/1	80/1	80/1.5
Layer 4	70/1	70/1	70/1.5
Layer 5	60/1	60/1	50/1.5
Layer 6	50/1	50/1	40/1.5
Layer 7	40/1	40/1	30/1.5
Layer 8	30/1	30/1	0/5
Layer 9	20/1	20/1	
Layer 10	10/1	10/1	
Layer 11	0/1	0/5	

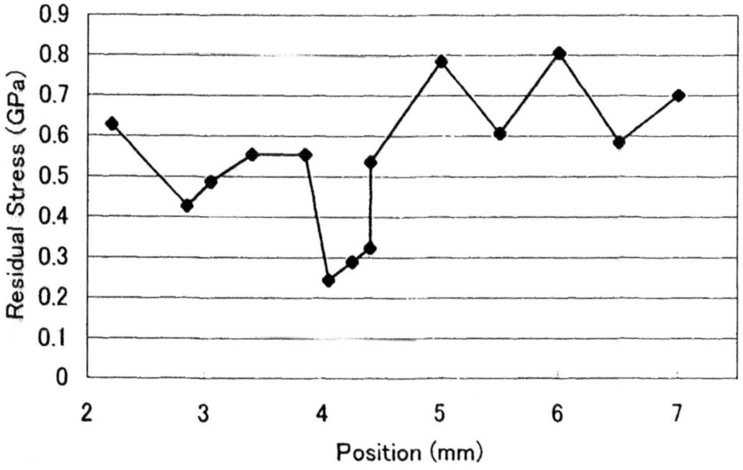

Figure 4.23. Stress distribution in FGM2 measured by Raman spectroscopy.

2.3.3 Fluorescent Measurement

It is well known that ruby shows a strong fluorescence at 694.3 and 692.9 nm, and those peaks shift with the applied load and temperatures. This phenomena is widely used to calibrate the pressures in high-pressure chambers such as Bridgman cells or diamond anvil cells. The R1 and R2 peaks shift 0.365 nm/GPa (7.6 cm^{-1}/GPa) and 0.0068 nm/K respectively [29]. Those shifts are larger than that of the ZrO_2 Raman peaks and are sufficient for measuring the residual stresses in the FGM. A treatment was made to add ruby powders to the ZrO_2 to fabricate the FGM. The concentrations of 1 and 0.1 wt% were tested and 0.1 wt% was found to be adequate for easily locating a ruby particle under the fluorescence microscope. The ruby fluorescence can be stimulated by the 488 nm laser beam which is used to obtain the Raman spectra. However, the fluorescence is too faint to be observed under a conventional microscope, so Mizuhara *et al.* used an ultraviolet illuminating unit. Since the UV light is more efficient than the 488 nm laser light, the same UV source was also used to stimulate the fluorescence.

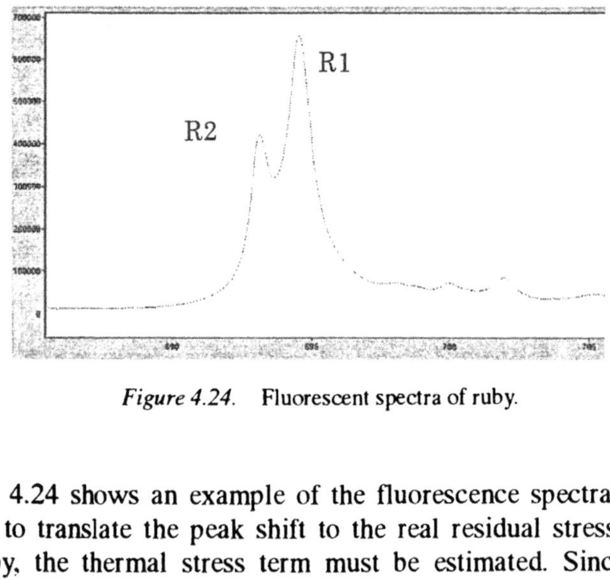

Figure 4.24. Fluorescent spectra of ruby.

Figure 4.24 shows an example of the fluorescence spectra of the ruby. However, to translate the peak shift to the real residual stress without the added ruby, the thermal stress term must be estimated. Since FGMs are sintered at high temperatures, and as shown in Table 4.3, the thermal expansion coefficient of ZrO_2 is twice as large as Al_2O_3, thus around 1.0 GPa compressive stress is expected. Also as shown in Figure 4.25, this thermal stress is relaxed at the surface. Figures 4.25 and 4.26 show the peak positions along the depth of a FGM1 and FGM2 respectively. As can be seen in the figures, the residual stresses are compressive and some tendencies along the depth were observed. At the top surface, the stress is low and it increased around the composition of 90% ZrO_2 and gradually reduced with the increasing depth. However, assuming that the surface of the 100% ZrO_2 layer is stress-free then the residual stresses are tensile at the lower ZrO_2 concentration layers. This is reasonable since the thermal expansion coefficient of stainless steel is twice as large as that of ZrO_2.

Comparing this result with Figure 4.23, the results obtained by Raman and fluorescence appear fairly consistent.

Table 4.3. Mechanical properties of zirconia and ruby.

	ZrO_2	Al_2O_3
Young's Modulus	210~ 220GPa	400Gpa
Poisson's ratio	0.25	0.25
Thermal expansion Coeff.	10.5x10^{-6} /K	4.9x10^{-6}/K

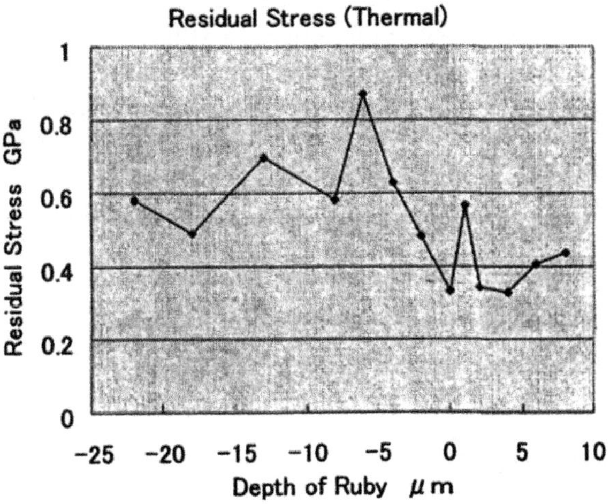

Figure 4.25. Stress distribution in FGM1.

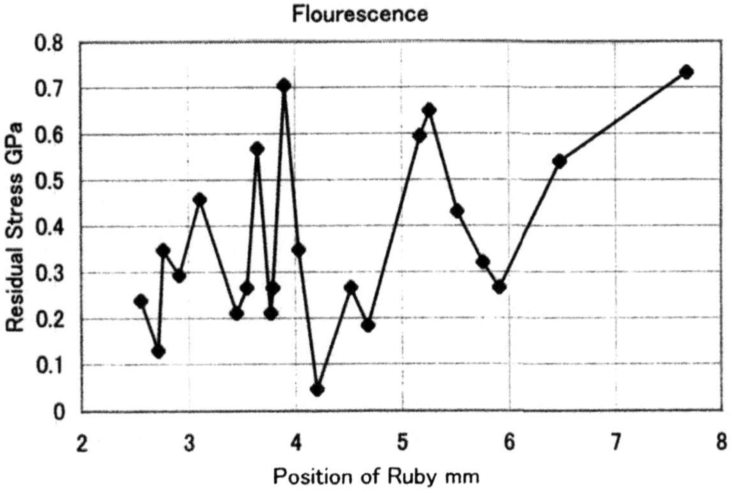

Figure 4.26. Stress distribution in FGM2.

The reason why scatter in the data is larger in the fluorescence method is the fluctuation in the thermal stress caused by the ruby itself. The other problem in the fluorescent method was the difficulty in finding the ruby particles at low ZrO_2 concentration layers. Since one kept the concentration of the ruby in the ZrO_2 constant at 0.1%, in the lower ZrO_2 concentration layers, the ruby concentrations are very low. However, in the lower ZrO_2 concentration region, less fracture is expected, so the residual stresses in these layers are not that important.

2.3.4 Conclusions

The feasibility of two methods for evaluating residual stresses in ZrO_2-based FGMs was evaluated and the following conclusions were reached.
(1) Residual stresses are namely tensile and increase with increasing depth.
(2) Both techniques could be applicable with some improvements.
(3) The Raman technique still needs more improvement in accuracy and repeatability.
(4) The fluorescence technique needs some mean for estimating the thermal stress during the process.

3. OXIDATION RESISTANCE

3.1 Introduction

Nb-Si alloys have good potential as new high-temperature structural materials because of their low density, high melting points and high strength at high temperatures. However, their oxidation resistance is extremely poor at high temperatures and there are two methods to improve this.

One method is adding some elements such as B, Cr and Al to the alloys. Meyer *et al.* reported that and addition of less than 2 wt% boron-doped Mo_5Si_3 produced extremely good oxidation resistance in the temperature range of 1073K to 1723K, because a non-porous thin protective borosilicate scale was formed [30, 31]. In addition, adding Cr and Al to the Nb-Si alloys is expected to improve the oxidation resistance. In the Nb-Si-X (X: B, Cr and Al) systems, several ternary inter-metallic compounds exist. These compound phases are formed in the third element doped alloys, although the oxidation behaviors and mechanical properties of these compounds have not yet been adequately investigated.

Another method is coating the Nb-Si alloys with good oxidation-resistant materials such as transition metal silicides. The compositions of niobium silicides are close to those of the Nb-Si alloys, although both Nb_5Si_3 and

$NbSi_2$ have inadequate oxidation resistance. It is therefore necessary to investigate the oxidation resistance and mechanical properties of the intermetallic compounds of the Nb-Si-X (X: B, Cr and Al) system.

In this work, powder compacts of the Nb-Si-X (X: B, Cr and Al) system were prepared by the spark plasma sintering method, and their microstructures and oxidation behaviors were investigated.

3.2 Preparation and Measurement

The compositions of the Nb-Si-B compacts investigated were Nb-37.5Si (Nb_5Si_3 matrix), Nb-34Si-10B ($Nb_5Si_3B_2$ matrix), Nb-66.7Si ($NbSi_2$ matrix) and Nb-66.7B (NbB_2 matrix) (at%). The compacts were prepared via the following procedures. Nb (99.8% pure, < 10 micron), Si (99.9% pure, –300 mesh) and B (99% pure, –300 mesh) powders were blended by a hexane-filled ball mill for 3.6 ks. The blended powders were spark plasma sintered at 1273K to 1473K under a pressure of 5 MPa for 0.3 ks, and then crushed and milled in a mortar and a hexane-filled ball mill. Compacts with dimensions 20 mm in diameter and 5 mm thickness were prepared by re-sintering the milled powders at 1773K to 2073K under a pressure of 49 MPa for 0.3 ks. Specimens (2x2x2 mm^3) were cut from the compacts, and their oxidation behavior was investigated by TGA.

The compositions of the Nb-Si-Cr compacts investigated were Nb-66.7Si-xCr (x = 0, 10, 20, 33.3 at%), while those of Nb-Si-Al compacts investigated were Nb-47Si-20Al ($Nb_3Si_5Al_2$ matrix) and Nb-25Si-15Al ($Nb_{10}Si_3Al_3$ matrix) (at%). These compacts were prepared via the same procedures as the Nb-Si-B compacts using Nb (99.8% pure, < 10 micron), Si (99.9% pure, –300 mesh), Cr (99.9% pure, –300 mesh) and Al (99.5% pure, –325 mesh) powders. The oxidation behaviors of the specimens (4x4x3 mm^3) cut from each compact was investigated by TGA.

3.3 Oxidation Behavior of SPSed Compacts

Figure 4.27 shows the mass change of the Nb-Si-B compacts exposed to dry air at 1523K. The mass change of the $Nb_5Si_3B_2$ matrix (Nb-34Si-10B) compact was smaller than that of the Nb_5Si_3 matrix (Nb-37.5Si) compact.

The oxidized layer formed on the $Nb_5Si_3B_2$ compact exposed for 18 ks did not spall, while that of the Nb_5Si_3 compact easily spalled. These results indicate that the oxidation resistance of the $Nb_5Si_3B_2$ matrix compact is superior to that of the Nb_5Si_3 matrix compact, although the oxidation resistance of the $Nb_5Si_3B_2$ matrix compact is extremely poorer than that of the $NbSi_2$ matrix (Nb-66.7Si) compact. The thickness of the oxidized layer formed on the $Nb_5Si_3B_2$ matrix compact exposed to dry air for 18 ks was about 250 micron by SEM observation. According to EDS and XRD

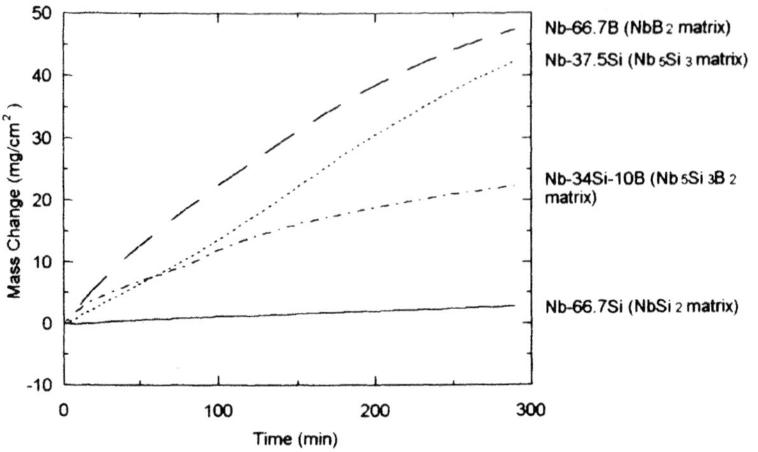

Figure 4.27. Continuous mass change of Nb-Si-B compacts exposed to dry air at 1523K.

analyses, the oxidized layer consisted of Nb_2O_5 and SiO_2 phases. When the Mo_5Si_3B compact was exposed to air at 1073K to 1723K, a thin protective borosilicate layer was formed by evaporation of MoO_3 [30]. On the other hand, the vapor pressure of Nb_2O_5 is very low at 1523K, which would prevent the formation of the protective borosilicate layer on the $Nb_5Si_3B_2$ compact.

Figure 4.28 shows continuous mass change of the Nb-66.7Si-xCr (x=0, 10, 20, 33.3 at%) compacts exposed to dry air at 1023K. At 1023K the $NbSi_2$ compact shows pest oxidation behavior [32], although the $NbSi_2$ compacts with more than 10 at% Cr added show extremely good oxidation resistance. SEM observation and EDS analysis showed a 1 to 3 micron thickness protective SiO_2 scale formed on the Cr doped compacts exposed to dry air at 1023K for 360 ks. However, their oxidation resistance became poor at 1573K because of the evaporation of CrO_3 [33].

The continuous mass change of the Nb-Si-Al compacts exposed to dry air at 1573K is shown in Figure 4.29. The mass change of the Nb-47Si-20Al ($Nb_3Si_5Al_2$ matrix) compact was much smaller than that of the other compacts. Figure 4.30 shows a cross section of the oxidized layer of the Nb-47Si-20Al compact exposed to dry air at 1473K for 360 ks. A thin protective Al_2O_3 layer formed on the compact exposed at 1373K to 1573K. Although the oxidation resistance of the Nb-47Si-20Al compact was poor at temperatures lower than 1273K, the compact shows extremely good oxidation resistance after being exposed to air at 1573K for 180 ks.

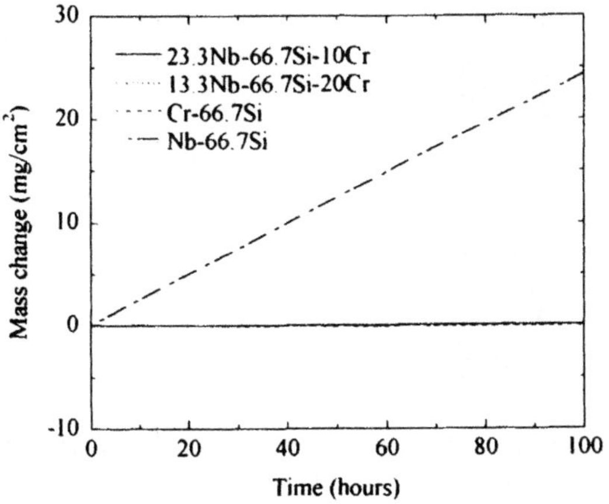

Figure 4.28. Continuous mass change of Nb-66.7Si-xCr (x=0, 10, 20, 33.3 at%) compacts exposed to dry air at 1023K.

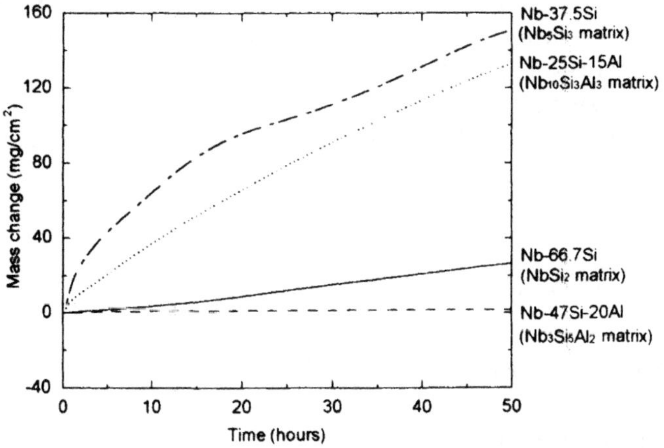

Figure 4.29. Continuous mass change of Nb-Si-Al compacts exposed to dry air at 1573K.

As described above, the Nb-47Si-20Al compact showed extremely good oxidation resistance at 1373K to 1573K, while all of the Nb-Si-B and Nb-Si-Cr compacts showed poor oxidation resistance at high temperatures. However, the mechanisms of the Al_2O_3 scale forming on the Nb-47Si-20Al compact are unclear and more work is needed to clarify this.

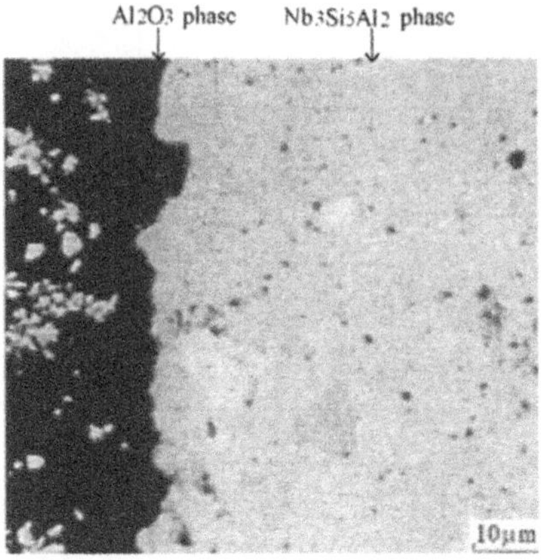

Figure 4.30. SEM photograph showing a cross section of the oxidized layer of Nb-47Si-20Al compact exposed to dry air at 1473K for 360 ks.

3.4 Conclusion

Powder compacts of the Nb-Si-X (X: B, Cr, Al) systems were prepared using the spark plasma sintering method, and their oxidation behavior at high temperatures was investigated. The oxidation resistance of the Nb-Si-B and Nb-Si-Cr compacts was poor at high temperatures. However, the Nb-47Si-20Al compact showed extremely good oxidation resistance at 1373K to 1573K, because of the formation of a thin protective Al_2O_3 scale.

4. MECHANICAL PROPERTIES

4.1 Fracture and Deformation

4.1.1 Introduction

A significant amount of theoretical work on the fracture and deformation of graded structures exists [34-40]. It has been revealed that the fracture resistance of a joint may increase or decrease when a graded structure is employed, depending on a variety of factors, including the crack orientation, the relative elastic and plastic property mismatches, the magnitude and orientation of thermal residual stresses, and the inherent fracture properties of the constituents. Unfortunately, there is very little experimental evidence to support the predicted behavior. It is the objective of this study to provide two examples in the fracture and deformation of ductile/brittle graded joints.

4.1.2 Preparation and Testing

Dense, graded composites either 25 mm or 50 mm in diameter and approximately 35 mm in height were formed by powder metallurgy methods in a vacuum hot press, as described in detail elsewhere [41-43]. Ni/Al_2O_3 composites consisted of monolithic Al_2O_3 on both ends of a cylinder, and monolithic Ni in the middle. Graded layers 1 mm in thickness of composition 80/60/40/20[†] separated monolithic Ni and Al_2O_3. Cu/W composites consisted of 80wt.% W composite material on the ends of the cylinder, and monolithic Cu in the middle. The ends and the cylinder center

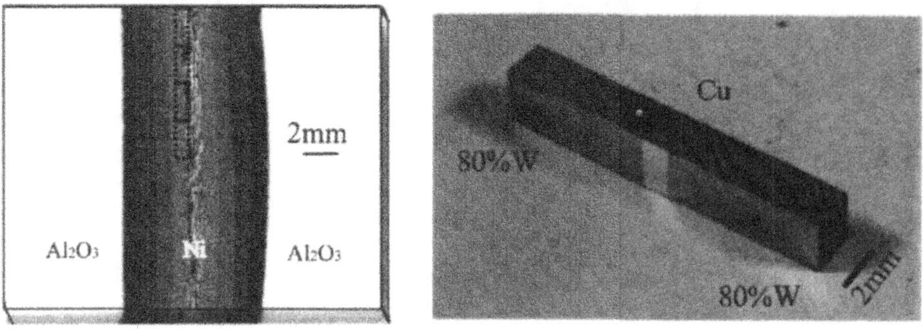

Figure 4.31. a) An optical micrograph of an incremental notch geometry for a Ni/Al_2O_3 graded composite is shown. The four rectangles drawn in the figure indicate the four sequential notch increments in which material was removed with a diamond saw; these notches existed within the 80%Ni layer. b) An optical micrograph showing the Cu/W geometry for four point bending.

were separated by graded layers 2 mm in thickness of composition 20/40/60 weight percent W. Samples of specific mechanical testing geometries were cut from the cylinders, as shown in Figure 4.31. Most of the mechanical testing reported here was conducted in an optical moiré interferometer. Phase shifting techniques were utilized to optimize the resolution, as described elsewhere [41, 44].

4.1.3 Displacement in Ductile/Brittle Joints

Each increment of material removal for the incremental notch sample (Figure 4.31(a)) resulted in displacement on the order of several microns as indicated in Figure 4.32. Only displacement positions for a single line scan across the top of the sample where the displacement was maximum are shown here. Most of the strain (dV/dy) occurs in the Ni-rich regions.

[†] The numbers correspond to weight percent Ni; remaining material is Al_2O_3.

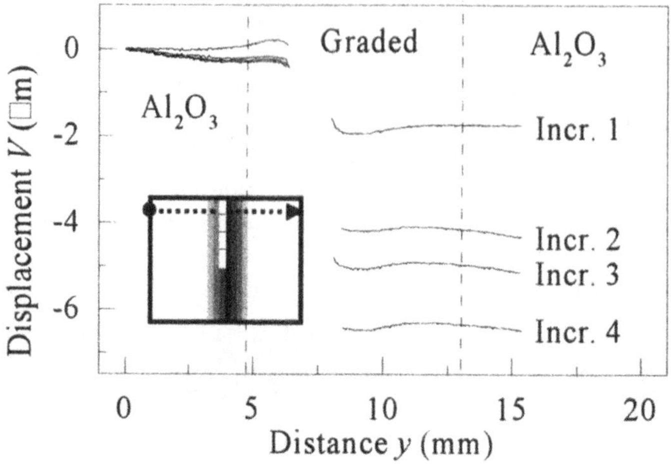

Figure 4.32. Displacement as a function of distance in the direction of the gradient for four notch increments. Each notch increment results in displacement due to residual stress relaxation.

Figure 4.33. illustrates the measured crack angles from two samples fractured in four point bending. Clearly, in both cases, the crack is deflected towards the copper-rich part of the graded composite. Figure 4.34. reveals a distortion in the local displacement field around the loaded notch tip, prior to fracture. It is noted that a homogeneous bar in bending would exhibit a symmetric displacement field around the notch tip. The shift in displacement field corresponds qualitatively to the shift in crack path observed in Figure 4.33.

4.1.4 Discussion

Figure 4.32 indicates that substantial compressive residual stress is relieved when material is removed. Further analysis [41] has revealed that the highest strain (e.g., the slope of the curve in Figure 4.32 is the transverse strain, ε_V = dV/dy) is in the nickel-rich compositions. Most of the strain must be elastic, since plasticity in these samples requires higher levels of strain [41]. On the other hand, the strains measured very close to the notch approach are much higher and probably exceed plastic strains.

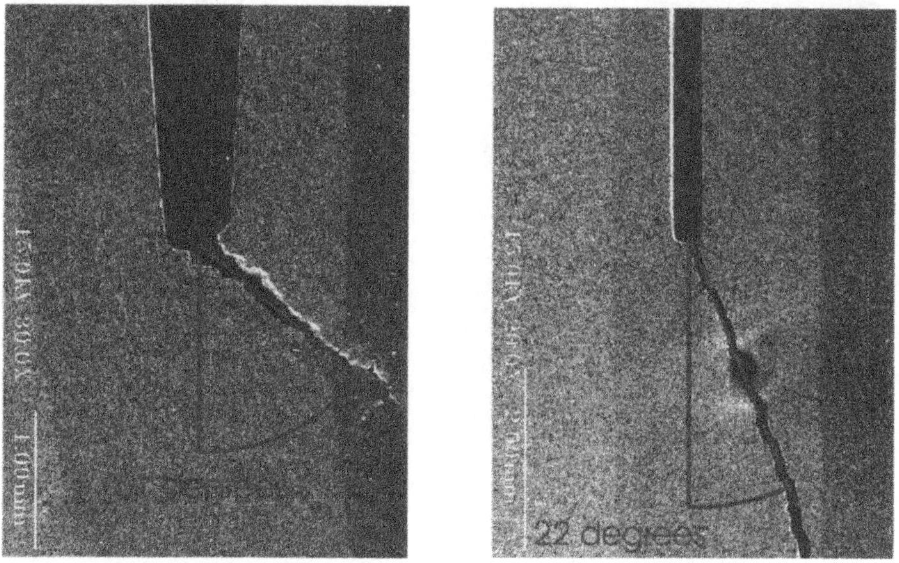

40 wt. % W Layer
a)

20 wt. % W Layer
b)

Figure 4.33. Scanning electron micrographs showing crack deflection angles in fractured Cu/W graded four point bending bars. a) 40 wt. % W layer was notched. b) 20 wt. % W layer was notched. An approximate measurement of the crack angle is indicated. In both figures, the pure copper layer is to the right. A scale bar is shown.

The fact that the deformation due to bending shown in Figures 4.33 and 4.34 is entirely elastic, as evidenced by unloading and reloading experiments, indicates that the elastic mismatch between the various composite layers has a large influence on the stress field which the beam experiences in bending. Finite element methods (FEM) simulating these Cu/W bend bars have also predicted displacement fields very similar to those observed in the PSMI results [45]. As shown in Figure 4.35, FEM predicts that the maximum tensile stress around the notch tip in a composite beam under load is inclined towards the copper rich layers. It is also apparent from Figure 4.35 that thermal residual stresses have a very small influence on shifting the notch tip stress fields. More extensive FEM analyses have shown that the effect of the notch on redistributing the applied stress field is very small or negligible. The fact that the effect of thermal residual stresses on the highest maximum principal stress direction are very small supports this latter conclusion; residual stress redistribution in the sample due to the presence of a notch most likely does not substantially change the highest maximum principal stress direction.

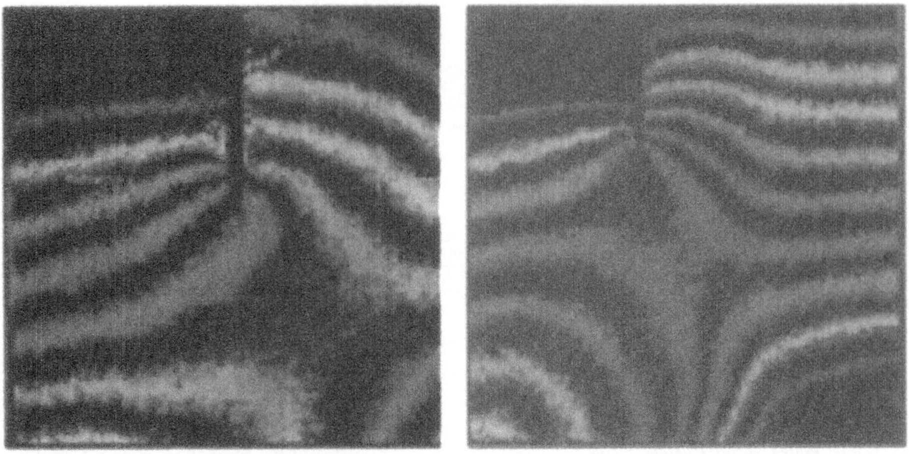

40 wt. % W Layer
a)

20 wt. % W Layer
b)

Figure 4.34. PSMI images of the bending samples in Figure 3, under load, prior to crack propagation. Both samples are under elastic loading. a) 40wt. % W. b) 20 wt. % W. The length of the notch in both figures is 2.5 mm.

4.1.5 Conclusions

Experimental measurements using PSMI have shown that the process of machining graded samples results in a stress redistribution. Most of the displacement for a notched bar occurs in the metal-rich regions of the sample. It does not appear that this stress redistribution significantly affects the crack trajectory in the case of Cu/W composites. Experimental and FEM results here indicate that elastic mismatch in material properties are the principal reason the stress field around the notch in the 20 and 40% W layers in Cu/W composites is asymmetric. It is likely that this asymmetric stress field causes a crack emanating from the notch to deflect at a prescribed angle, as observed experimentally (Figure 4.33).

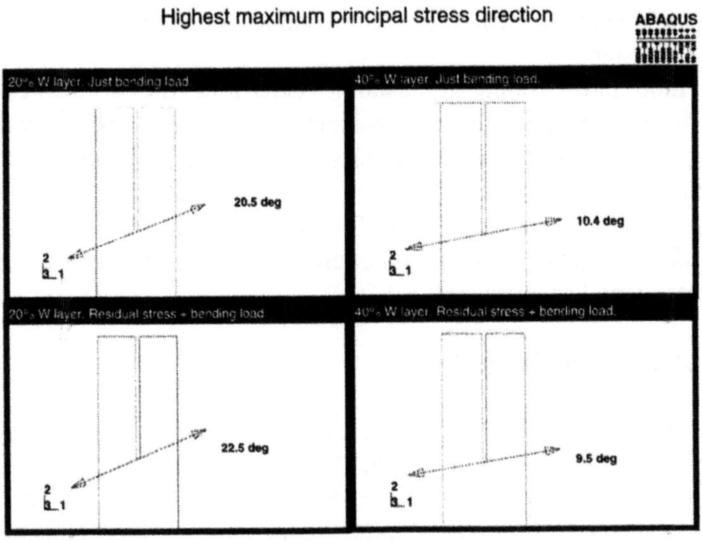

Figure 4.35. FEM simulations indicating the region of highest maximum principal stress for four situations. The top two figures illustrate results for notches in the 20 and 40wt. % W layer in bending. The bottom two figures give results for the same notches in the top two, but including the effect of thermally induced residual stresses which arise from thermal expansion coefficient mismatch between Cu and W.

4.2 Instrumented Indentation

4.2.1 Introduction

The use of thin films and coatings is becoming an increasingly important part of component design. Some films, for example in wear and tribology applications, are used specifically to improve the mechanical properties of a surface. However, even when the films are incorporated primarily for their electronic, magnetic, optical or thermal properties, some knowledge of their mechanical properties is often valuable. For example, a ceramic coating applied as a thermal barrier may not experience any direct mechanical loading, but knowledge of its elastic moduli is important for the calculation of stresses that will develop in it during thermal cycling.

When the mechanical properties of thick coatings are needed, it is sometimes possible to prepare freestanding pieces of coating material that can then be tested as if they were bulk material, but the specimens are often delicate and difficult to prepare. In some cases there may be questions about whether the freestanding coating is really the same material as the coating used in service, perhaps because the freestanding coating was initially deposited on a sacrificial substrate material different from that used in the actual application. For thinner coatings and films (< 10 μm thick), testing of a freestanding film is nearly always impossible, and tests must be performed on a film/substrate system.

The test method most commonly used to determine the mechanical properties of thin films on substrates is instrumented indentation, also known as depth-sensing indentation, recording hardness or, when the test forces are very small, nanoindentation. In this section, the basic technique will be described, along with its application to thin film and coating measurements. Results from spark plasma sintered functionally graded metal-ceramic coatings will be presented as an example.

4.2.2 Description of the Technique

Instrumented indentation testing (IIT) is similar to conventional micro-hardness testing, in that an indenter is loaded onto a surface and then removed. The indenter is almost always made of diamond that has been ground and polished in one of several geometries, the most common being a sharp three-sided pyramid (Berkovich), a four-sided pyramid (Vickers) or cones with spherical tips of various radii. IIT differs from microhardness testing, however, in several critical ways. First, the force on, and displacement of, the indenter is monitored and recorded continuously throughout the loading-unloading cycle. Second, the result of the indentation experiment comes not from a measurement of the residual impression on the test piece, but from an analysis of the recorded force-displacement data. This analysis permits the determination not only of the hardness of the test piece, but also its elastic modulus. It should be noted that the elastic modulus determined with IIT is not exactly equal to Young's modulus. The elastic modulus determined from IIT, often termed the 'indentation modulus,' E_i, is equal to $E/(1-\nu^2)$, where E is Young's modulus and ν is Poisson's ratio[46]. Poisson's ratio is not determined independently in IIT.

The use of force-displacement data, as opposed to a size measurement of a residual impression, to obtain hardness and other mechanical properties is the key to the technique's value for thin film and coating characterization, since in general it is necessary to keep the indentation (and most of the stress field supporting it) within the film in order to obtain the properties of the film. In principle, the only limit on how small a volume of material can be probed comes from the accuracy and resolution of the force and displacement measurement systems. In practice, there are commercial IIT systems capable of obtaining usable force-displacement data for indentations less than 10 nm deep. However, a proper quantitative analysis of those data requires knowledge of the contact area between the tip and specimen, which in turn requires knowledge of the tip geometry. This is not easy to determine for the first 10B20 nm of a tip.

4.2.3 Application of IIT to Spark Plasma Sintered Graded Systems

Spark plasma sintering (SPS) is a method for rapidly sintering a wide range of materials and a wide range of material combinations [47]. Typically, powders are packed into a cylindrical bore in graphite die, uniaxial pressure is applied by cylindrical punches entering the top and bottom of the die opening, and the powder is sintered by the application of pulses of high electric current through the powder and die. These current pulses create a plasma in the voids between powder particles that results in rapid heating, compaction and sintering of the powder. Functionally graded materials (FGMs) can be produced by loading the die with layers of powder mixtures where the volume fraction of each component varies from one layer to the next.

Several graded metal-ceramic coating systems were tested using IIT. All specimens were prepared by the Sumitomo Coal Mining Group of Izumi Technology Company, and were studied at NIST as part of the Civil Industrial Technologies Arrangement between the U.S. Technology Administration and the Japanese Agency of Industrial Science and Technology. All specimens consisted of a pure metal as substrate material, mixed metal-ceramic interlayers, and a pure ceramic top-coat. Systems ranged in complexity from having only two mixed interlayers (0.50 and 0.75 volume fraction ceramic) to having nine interlayers. After sintering, each mixed layer was on the order of 0.5 mm thick, while the ceramic topcoats were typically 1.5 mm thick and the metal substrate layers were several millimeters thick. Occasionally, interlayer thickness was varied above or below the 0.5 mm average to more closely control the concentration gradient. All specimens studied in this work were produced as cylinders of 20 mm, 30 mm or 40 mm diameter, with the metal-ceramic gradient parallel to the cylinder axis.

A variety of indenter geometries and applied forces were used to test different properties of the SPS FGMs. The current report will focus entirely on data obtained with a sharp diamond indenter (a three-sided diamond pyramid known as a Berkovich indenter) at forces of 1 N or 2 N. The Berkovich indenter was chosen because the resulting data could be analyzed for both hardness and elastic modulus. Loads were chosen such that the indentations were large enough to obtain average properties in the heterogeneous mixed regions, yet small enough that the indentations did not cause significant cracking in the specimens. Such cracking can cause errors in the determination of material properties.

In order to study the mechanical properties of the coatings through their thickness, and in some cases laterally within layers as well, cross-sections were prepared by cutting and polishing surfaces containing each specimen's original central axis. Rows of indentations were made parallel to the concentration gradient, typically with 100 μm spacing between indentations in each row.

The first SPS graded coating system studied with IIT was Ti/Al_2O_3 with 4 layers: pure Ti substrate; 0.50 volume fraction Al_2O_3; 0.75 volume fraction Al_2O_3 and pure Al_2O_3 topcoat. All indentations were made along the center axis of each specimen. Three specimens having only minor variations in sintering heating rates and times were studied. Of those, only one achieved values of modulus and hardness in the alumina topcoat comparable to those for fully dense polycrystalline alumina sintered by conventional means. Although significant porosity was not visible in the polished cross-sections of any of the three specimens, IIT clearly revealed incomplete sintering in two of the three specimens.

The second coating system studied consisted of several sets of graded ZrO_2/410 stainless steel coatings. The first set had 6 mixed layers: 0.30, 0.50, 0.60, 0.70, 0.80 and 0.90 volume fraction ZrO_2, plus a pure steel substrate layer and a pure ZrO_2 topcoat. The zirconia was partially stabilized with a 0.03 mass fraction of yttria. The key observation made in this first set of graded zirconia coatings was that there were sharp discontinuities (drops) in both the hardness and modulus at some of the internal interfaces, particularly the interface between the last mixed layer (0.90 ZrO_2) and the pure zirconia topcoat. Since this drop suggested the possibility of a weak interface, additional mixed layers with 0.94, 0.95 and 0.97 volume fraction ZrO_2 were added in a second set of coatings, to help smooth the transition to pure zirconia.

The modulus and hardness results for one of the new coatings are shown in Figure 4.36. Each of the nine mixed layers is approximately 0.5 mm thick, with the interface between the steel substrate and the 0.30 ZrO_2 layer at an axial position of approximately 1mm. Although there is no significant discontinuity in either the hardness or modulus at interfaces near the ZrO_2 topcoat (axial positions greater than 6 mm), there is a large drop in the modulus in going from the pure steel substrate to the first mixed layer. This anomalously low modulus persists through the 0.50, 0.60 and 0.70 volume fraction ZrO_2 layers. The hardness shows only a slight drop in the same region. The multiple curves in the figure denote measurements taken at different radial distances from the central axis; no significant differences in properties are observed as a function of radial position, indicating that the in-plane uniformity of each layer is good.

The well-behaved, essentially monotonic increase in hardness observed with increasing ZrO_2 content is expected. However, the low value of modulus observed for intermediate ZrO_2 concentrations is not yet understood, and is certainly not explained by any type of law of mixtures analysis, as the moduli for the two individual components are nearly identical. It is possible

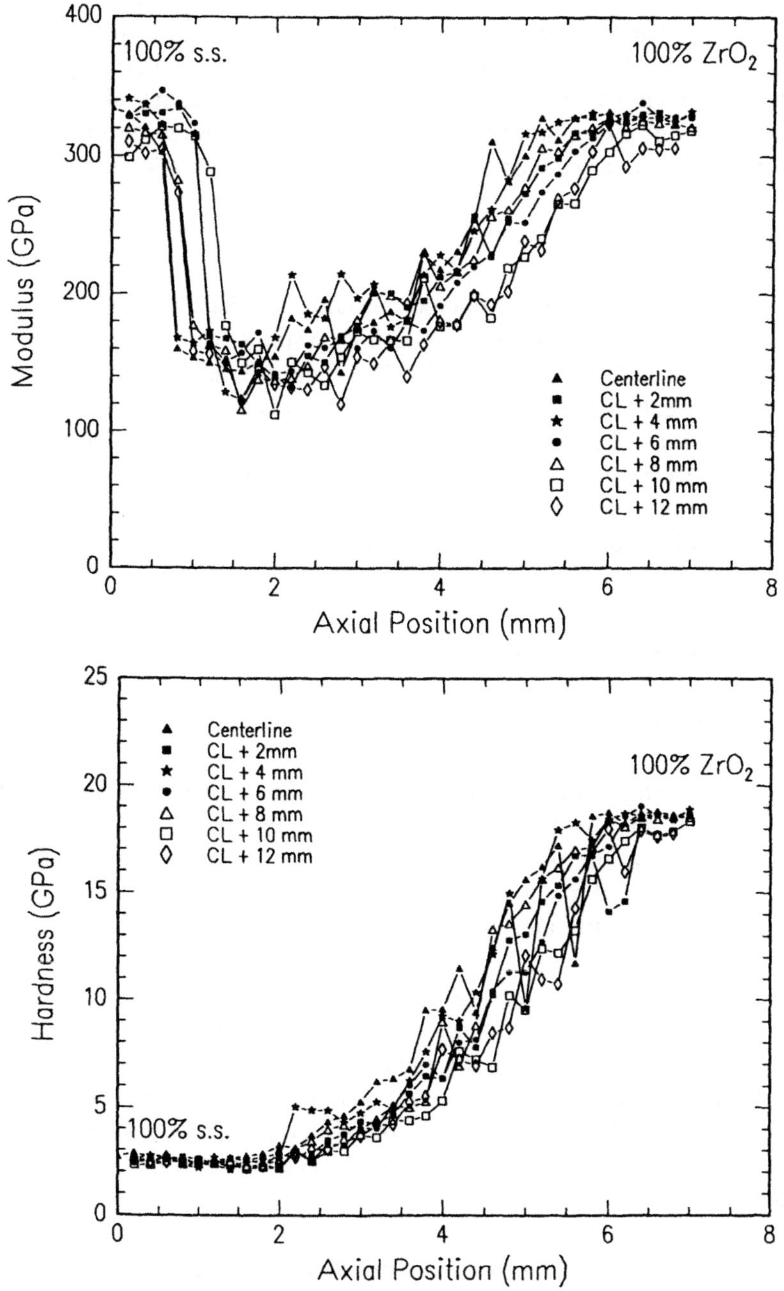

Figure 4.36. Indentation modulus and hardness for a graded ZrO_2/410 stainless steel coating measured across the gradient both on the specimen center axis and at various radial distances from the center axis. The full radius of the specimen was 15 mm. Little variation is seen as a function of radial distance. The axial position of 1 mm, where the modulus drops abruptly, corresponds to the interface between the pure steel substrate and the first mixed layer, which contains 0.30 volume fraction of ZrO_2.

a significant drop in hardness as well as modulus, as was observed in the Al_2O_3/Ti system described above but is not seen here. However, the same qualitative effect of a reduced modulus in the intermediate region on the same specimen has been observed by Takatsubo using ultrasonic wave propagation [48], he also observed some porosity in these middle layers, but the amount has not yet been quantified. This effect warrants further study.

4.2.4 Conclusions

Local measurements of elastic modulus and hardness of several functionally graded metal-ceramic coating systems have been made using instrumented indentation methods. The measurements have revealed both expected effects, such the relatively well-behaved increase in hardness with ceramic volume fraction, and surprises, such as anomalously low moduli in some metal-ceramic mixed regions. It has also provided information on the uniformity of properties within layers, which was good in all specimens studied, and has detected inadequate sintering in some specimens.

4.3 Properties of Large-Size FGMs

4.3.1 Introduction

Spark plasma sintering (SPS) is a pulse electric current sintering process that has recently been attracting attention. It has excellent features that enable the creation of new materials such as composite ceramics and functionally graded materials (FGMs) that relax stress by gradually changing the layer composition and characteristics from a ceramic layer to a metallic layer.

In this work, large-size heat-resistant ceramic/metal bulk FGMs (FGM/LB) of the thermal stress relaxed type that are suitable for applications as high-efficiency engine materials were fabricated. Observations were made of the microstructures of the sintered materials, their hardness distribution characteristics were evaluated and the tribological characteristics resulting from the composite implementation of their ceramic surface layer were examined.

4.3.2 Preparation, Observation, and Testing

Using ZrO_2 (3Y) powder containing 3 mol% of Y_2O_3 (with a specific surface area of 11.2 m^2/g by BET method) for the ceramic material and SUS410L powders (average grain diameters of 9μm and 60μm) for the metallic material, large bulk functionally graded materials (FGM/LB) in that this intermediate region does not sinter well, but one would then expect diameters of 100 mm and 150 mm were fabricated using an SPS machine. The temperature gradient sintering process was used with an irregularly shaped graphite die/punch shown in Figure 4.37. The intermediate mixed-graded layers have a ZrO_2 (3Y) volume fractions of 90, 80, 70, 60, 50, 40, 30, 20 and 10 mol% respectively while other parts are occupied by the ZrO_2 (3Y) and stainless steel powders. Table 4.4 shows the compositions and sintering conditions of the graded layers. The structures of the layer cross-sections of the FGM/LB sintered specimens were observed using scanning electron microscopy (SEM) and optical microscopy and their hardness distribution was also measured using a Vickers micro-hardness tester. Figure 4.38 shows the measurement locations on an FGM/LB sintered compact.

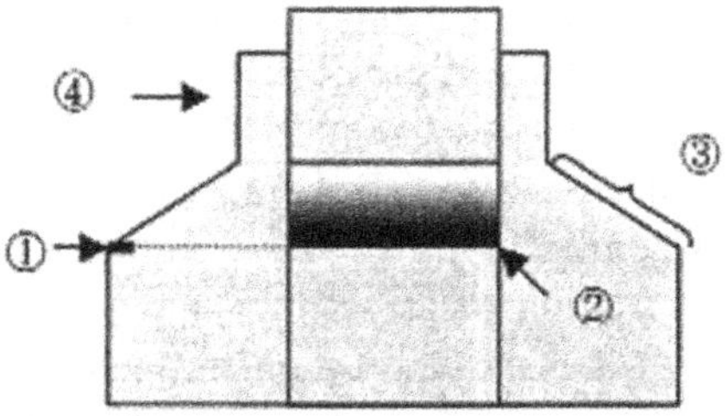

①SPS temperature (low-temperature section, using a thermocouple)
②SUS410L stainless steel
③Temperature gradient section
④Ancillary temperature gauge (high-temperature section, using an IR radiation thermometer)

Figure 4.37. Shape of temperature gradient graphite die and punch.

Table 4.4. Composition and sintering condition of graded layers.

Diameter	Stainless steel average particle size (μ m)						
	SUS layer	Intermediate layer	Number of intermediate layers	Thickness of 50% layer	Sintering temperature K	Pressure MPa	Specimen Name
100mm	60	60	9	3	1198	12	FY1997
	9	60	9	3	1198	19	FGM22
	60	60	9	3	1248	19	FGM13
	9	60	9	1	1248	19	FGM18
	60	60	9	1	1223	19	FGM14
	9	60	9	1	1223	19	981029
	9	9	9	1	1223	19	981202
	60	60	6	1	1223	19	981023
150mm	60	60	9	1	1213	8.5	981112
	9	60	9	1	1213	8.5	981225

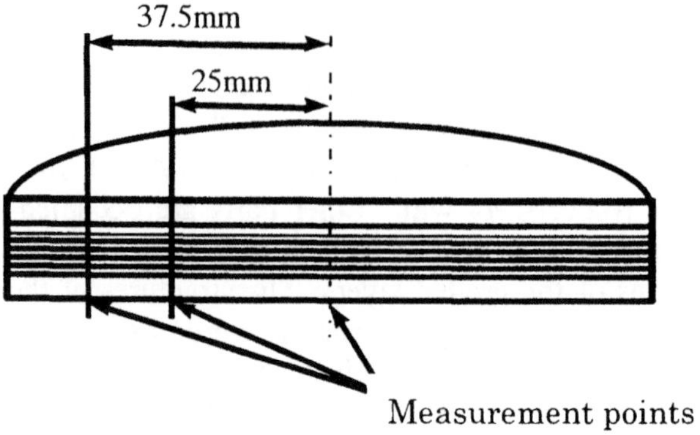

Figure 4.38. Vickers microhardness measurement points on a FGM.

To evaluate the tribological characteristics of the composite FGM, two different test pieces were sintered using ZrO_2 (3Y) + 20 wt% Al_2O_3 powder, prepared by mixing 20 wt% Al_2O_3 in ZrO_2 (3Y), and alternatively using ZrO_2 (3Y) + 20 wt% Al_2O_3 + 7 wt% $BaCrO_4$. These were obtained by adding the solid lubricant $BaCrO_4$ to the 100% ceramic layers of the powders described above. For friction wear testing, a pin-on-block type reciprocal high-temperature wear tester based on high-frequency induction heating was used to measure the friction coefficients of the samples, and the wear traces were observed using surface roughness measuring equipment.

4.3.3 Structural Observations

The pores which were observed in layers with the ZrO_2(3Y) content of 80% to 60% fabricated in FY 1997's 22 (specimen number of 1998/99) FGM specimens were not observed in the ZrO_2(3Y) 80% layer of FGM specimens sintered in 1998/99 at a higher pressure, or in any of the ZrO_2(3Y) 80% to 60% layers of FGM specimens sintered at a higher pressure and temperature. This provides evidence of the improvements achieved by increasing pressures and temperatures.

In the intermediate layers, stainless steel grains flattened in the direction perpendicular to the direction of the load during sintering were observed. Similar tendencies were also observed with the sintered FGM using micro-fine grains with an average grain diameter of 9μm.

4.3.4 Hardness Distribution Measurement

In the ZrO_2(3Y) 0% to 30% layers, the hardness in the central area was relatively lower than closer to the perimeter. When the 9μm fine grains were used as the stainless steel starting powder, improvements in the hardness and in the uniformity of the hardness distribution were observed in layers with large ZrO_2(3Y) content.

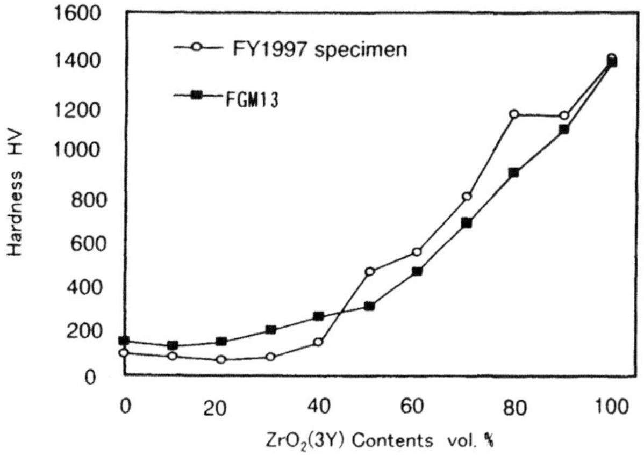

Figure 4.39. Vickers microhardness distribution of specimens with various sintering Temperatures and pressures.

With FGM13, with which both the pressure and temperature were increased, the hardness was lower than in the FY 1997 specimen in layers with a large ZrO_2(3Y) content but higher in the layers with a high ZrO_2(3Y) content as shown in Figure 4.39.

4.3.5 Tribological Characteristics

Figure 4.40 shows the temperature dependence of the friction coefficient of the ZrO_2 (3Y) + 20 wt% Al_2O_3 specimen, and Figure 4.41 shows that of the ZrO_2 (3Y) + 20 wt% Al_2O_3 + 7 wt% $BaCrO_4$ specimen. When the alumina (Al_2O_3) was added, the friction coefficient at near room temperature in the initial stage of sintering was about 0.65, it increased gently with the temperature rises later, decreased momentarily in the middle, reached its maximum values of nearly 0.8 at around 873K, and then decreased, with its minimum value of 0.55 at 1273K. When the solid lubricant ($BaCrO_4$) was added, the friction coefficient at near room temperature was about 0.5, it increased as the temperature rises later, decreased momentarily in the middle, reached its maximum value of nearly 0.8 at around 773K to 873K, and then decreased again. During falls in temperature, the changes in the coefficients of friction were less and the maximum values were lower than during temperature rises. Overall, the friction coefficient was lower with the specimen to which $BaCrO_4$ had been added.

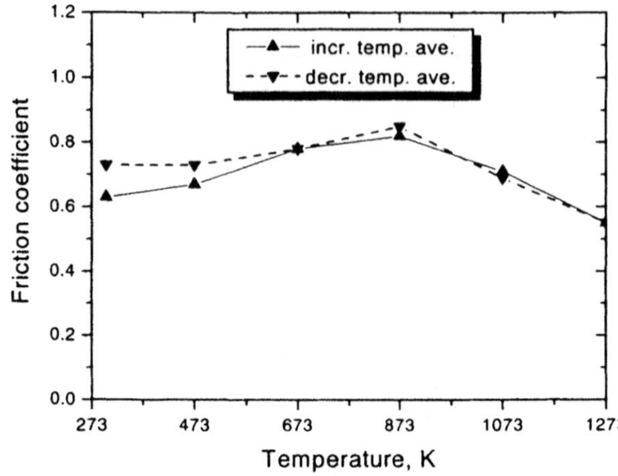

Figure 4.40. Temperature dependence of friction coefficient of ZrO_2(3Y) +20wt%Al_2O_3 composite sintered by SPS.

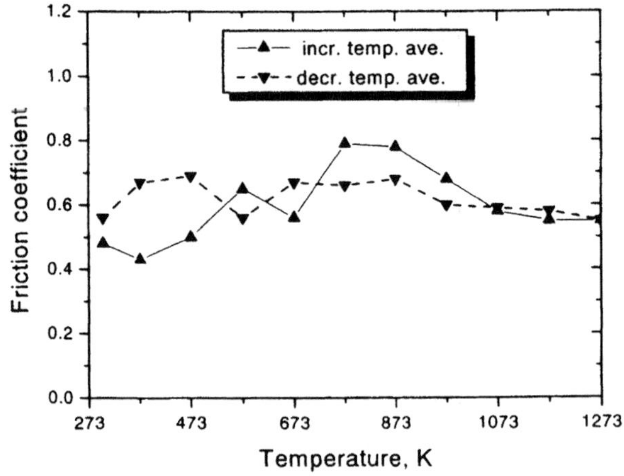

Figure 4.41. Temperature dependence of friction coefficients of ZrO_2(3Y) +20wt%Al_2O_3 +7wt%$BaCrO_4$ composite sintered by SPS.

Figure 4.42 shows a comparison of the friction trace profiles on the block sides of the specimens after friction wear testing. The addition of $BaCrO_4$ reduced the degree of friction about 2.5 times that of the specimen without it. This is regarded as occurring because the addition of $BaCrO_4$ reduced the sintering properties and made sufficient compaction impossible if the same temperatures were used. In other words, although the addition of the solid lubricant improves friction characteristics, it also deteriorates the sintering properties and wear characteristics.

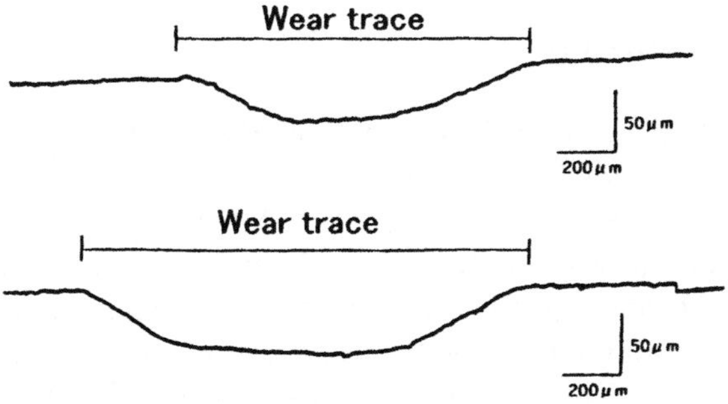

Upper figure : ZrO_2 (3Y)+ 20wt% Al_2O_3 Composite Ceramic
Lower figure : ZrO_2 (3Y)+ 20wt% Al_2O_3 + 7wt% $BaCrO_4$ Composite Ceramic

Figure 4.42. Comparison of wear trace profiles on block sides after reciprocal friction wear testing.

4.3.6 Summary

(1) To identify the effects of SPS related to size and shape in dealing with the trend of large bulk implementation of FGM/LB, various starting powder particle sizes, layer thicknesses and their combinations were employed to survey and test their effects on the sintering of φ100 mm and φ150 mm specimens. It was confirmed that non-uniformity increases as the specimen size increases. However, it was also found that the use of stainless steel starting powder material of a specified degree of fineness could improve the uniformity of the hardness and the hardness distribution of the ceramic-rich side while greatly reducing the residual pores in the interface areas between the metallic and ceramic particles. It was also found that with SPS, which is greatly dependent on the pressure used, it is necessary to increase the pressure as the sinter size increases and improve the pressure distribution to optimize the sintering process.

(2) With regard to improvements in tribological characteristics, it became clear that composite ceramic materials to which a $BaCrO_4$ solid lubricant have been added can be fabricated and are applicable for high-temperature domains up to 1273K. However, as there is a correlation between improvements in the friction coefficient and some deterioration in wear resistance, more studies are required into the dispersal quantities and the type of solid lubricant used.

4.4 Tensile Properties

4.4.1 Introduction

A NEDO Project called *Precompetitive Processing and Characterization of Functionally Graded Materials* was conducted as a 4-year joint research between the US National Institute of Standards and Technology, the Mechanical Engineering Laboratory, Sumitomo Coal Mining Co. Ltd., Group Izumi Technology Co. Ltd. and Yanmar Diesel Engine Co. in Japan from FY '96 through FY '99 to produce and evaluate new functionally graded materials (FGMs) using a spark plasma sintering technique. To evaluate the mechanical, physical and tribological properties of large-scale bulk FGMs, composed of a 3 wt% yttria stabilized zirconia (YSZ) layer, mixed layers of the zirconia and SUS410L (similar to AISI410) stainless steel and the stainless steel layer, various characteristics such as Vickers hardness, elastic modulus, tensile strength, friction coefficient, residual stress, lattice parameter were measured by various kinds of testing methods such as Vickers harness measurement, micro-indentation, tensile test, friction test, ultrasonic imaging test, X-ray diffraction, and Micro-Raman spectroscopy.

The tensile properties are among the most important factors in evaluating the performance of the FGMs. An FGM composed of zirconia/stainless steel was proposed for engine components, because the zirconia has high heat-resistance and high wear-resistance and the stainless steel has high mechanical strength. In this study, an investigation was conducted on the tensile properties of a large-scale bulk FGM composed of a 3 wt% yttria stabilized zirconia layer, a mixed layer of zirconia and SUS410L (similar to AISI410) stainless steel and a stainless steel layer.

4.4.2 Preparation and Testing

An 11 layer yttria-stabilized zirconia/SUS410L stainless steel type FGM 50 mm in diameter, composed of 100%YSZ, 90%YSZ/10%410Lss, 80%YSZ/20%410Lss,70%YSZ/30%410Lss,60%YSZ/40%410Lss,50%YSZ/50%410Lss,40%YSZ/60%410Lss,30%YSZ/70%410Lss,20%YSZ/80%410Lss, 10%YSZ/90%410Lss, and 100%410Lss layers was produced by spark plasma sintering. The thicknesses of the YSZ, mixed and 410Lss layers were 4, 1 and 4 mm, respectively. Two types of stainless steel powders 9 and 60 μm were used for the spark plasma sintering experiment. The sintering conditions were as follows. The applied pressures were 200, 300, and 400 kg/cm^2. The sintering temperatures were 970 and 1020°C. Two kinds of temperature control programs were used as follows. (1) The temperature was increased from room temperature to 920°C over a period of 28 min, then increased from 920 to 970°C over 2 min, and held at 970°C for 3 min. (2) The temperature was increased from room temperature to 970°C over a period of 30 min, increased from 970 to 1020°C over 2 min and held at 1020°C for 3 min. The tensile test pieces were cut from the composition-graded direction of the FGM by a diamond cutter and polished. The total length was 17 mm and the gauge length was 11 mm. The tensile testing was carried out at room temperature and at a tensile speed of 0.5 mm/min using a small tensile test machine. Microstructural observation was done on the fractured surface of the FGM.

4.4.3 Apparent Density and Tensile Strength

The apparent density of the 50-mm diameter zirconia/stainless steel type FGM was calculated on the basis of the weight and the measured size of the consolidated FGM. The apparent density of the fully-consolidated FGM was 6.98 g/cm^3.

Figure 4.43 shows the relationships of the applied pressure, P, the apparent density, D, and the sintering temperature in the 50-mm diameter zirconia/stainless steel type FGM containing SUS410L stainless steel powders 9 mm in diameter. It was found that the apparent density increased from 6.40 to 6.66 g/cm^3 by increasing the pressure from 200 to 400 kgf/cm^2 at a sintering temperature of 970°C. It was also found that the apparent density

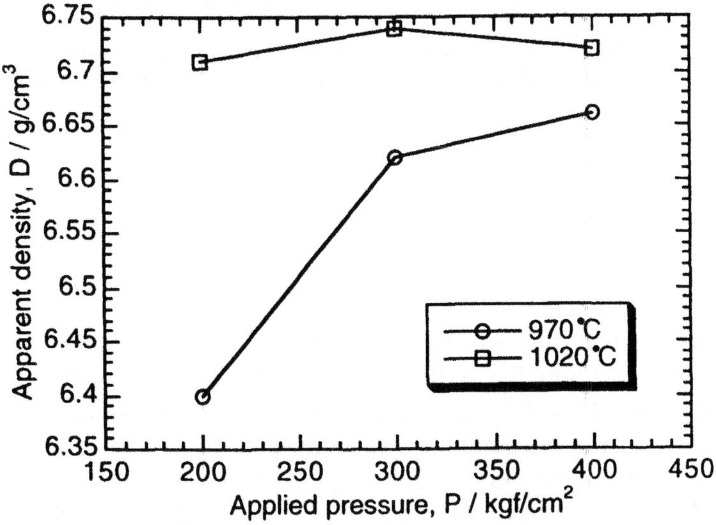

Figure 4.43. Relationships of applied pressure, apparent density and sintering temperature in a 50-mm diameter zirconia/stainless steel type FGM containing SUS410L stainless steel powders 9 μm in diameter.

was increased from 6.71 to 6.74 g/cm^3 and was decreased to 6.74 g/cm^3 by increasing the pressure from 200 to 400 kgf/cm^2 at a sintering temperature of 1020°C.

Figure 4.44 shows the relationships of the applied pressure, P, the apparent density, D, and the sintering temperature in the 50-mm diameter zirconia/stainless steel type FGM containing SUS410L stainless steel powders 60 mm in diameter. It was found that the apparent density was increased from 6.37 to 6.62 g/cm^3 by increasing the pressure from 200 to 400 kgf/cm^2 at a sintering temperature of 970°C. It was also found that the apparent density was increased from 6.56 to 6.80 g/cm^3 by increasing the pressure from 200 to 400 kgf/cm^2 at a sintering temperature of 1020°C.

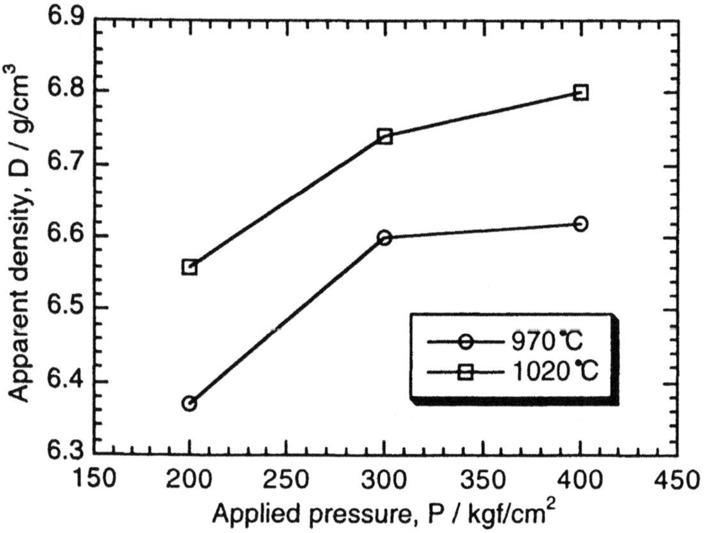

Figure 4.44. Relationships of applied pressure, apparent density and sintering temperature in a 50-mm diameter zirconia/stainless steel type FGM containing SUS410L stainless steel powders 60 μm in diameter.

It was concluded that the apparent density of the FGM containing SUS410L stainless steel powders 9 mm in diameter was almost equal to that of the FGM containing SUS410L stainless steel powders 60 mm in diameter.

Figure 4.45 shows the relationships of the applied pressure, P, the tensile strength, UTS, and the sintering temperature in the 50-mm diameter zirconia/stainless steel type FGM containing SUS410L stainless steel powders 9 mm in diameter. It was found that the tensile strength was increased from 61.9 to 110 MPa and then slightly decreased to 104 MPa by increasing the pressure from 200 to 400 kg/cm^2 at a sintering temperature of 970°C. It was also found that the tensile strength was increased from 30.5 to 109 MPa by increasing the pressure from 200 to 400 kgf/cm^2 at a sintering temperature of

1020°C.

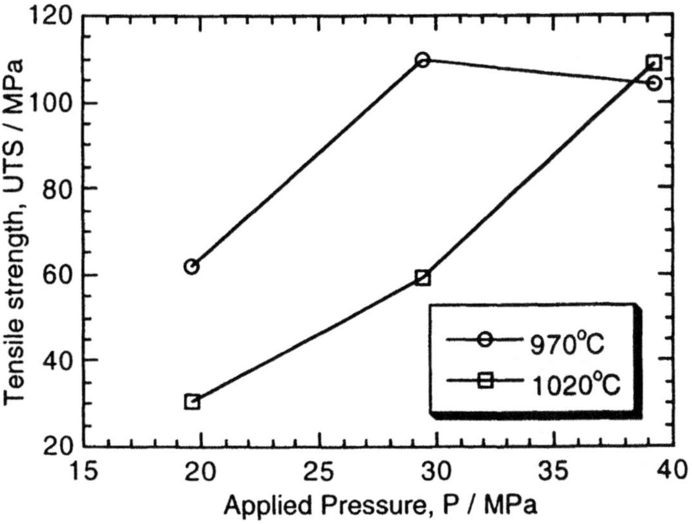

Figure 4.45. Relationships of applied pressure, tensile strength, and sintering temperature in a 50-mm diameter zirconia/stainless steel type FGM containing SUS410L stainless steel powders 9 μm in diameter.

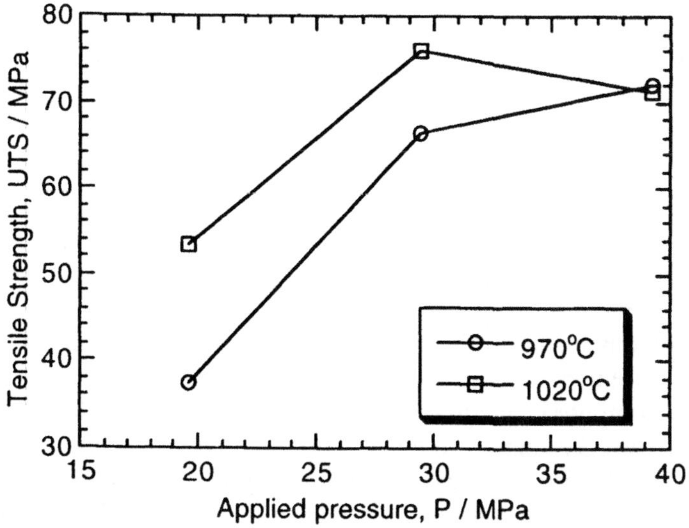

Figure 4.46. Relationships of applied pressure, tensile strength, and sintering temperature in a 50-mm diameter zirconia/stainless steel type FGM containing SUS410L stainless steel powders 60 μm in diameter.

Figure 4.46 shows the relationships of the applied pressure, P, the tensile strength, UTS, and the sintering temperature in the 50-mm diameter zirconia/stainless steel type FGM containing SUS410L stainless steel powders 60 mm in diameter. It was found that the tensile strength was increased from 37.5 to 71.9 MPa by increasing the pressure from 200 to 400 kgf/cm^2 at a sintering temperature of 970°C. It was also found that the tensile strength was increased from 53.4 to 76.1 MPa and slightly decreased to 71.3 MPa by increasing the pressure from 200 to 400 kgf/cm^2 at a sintering temperature of 1020°C. In general, it was found that the tensile strength in the 50-mm diameter zirconia/stainless steel type FGM containing SUS 410L stainless steel powders 9 mm in diameter was higher than that in the FGM containing SUS 410L stainless steel powders 60 mm in diameter.

4.4.4 Summary

It was concluded that the maximum tensile strengths in the 50-mm diameter eleven-layer yttria-stabilized zirconia/SUS410L stainless steel type FGM containing SUS410L stainless steel powders 9 mm in diameter were 110 and 109 MPa under the conditions of applied pressure of 300 and 400 kgf/cm^2 at sintering temperatures of 970 and 1020°C, respectively.

5. ELECTRICAL PROPERTIES

5.1 Functionally Graded Electronic Materials

5.1.1 Introduction

Electronic ceramics provide the basic components for the support of a variety of electronic products, including computers, industrial controls, consumer automotive devices and digital switches. Electronic ceramics can be used as passive components such as capacitors, resistors and inductors. For these components, chip size is becoming smaller and smaller. Ceramic capacitors are used in small-capacitance applications in consumer electronics, personal computers, peripherals and microprocessors. The size of these capacitors was 3.2 mm × 1.6 mm (3216 type) ten years ago, but the 1.0 mm × 0.8 mm (1008 type) size is now available. Soft ferrites are used as magnetic materials in transformers, inductors, recording heads and memory devices. Downsizing is also required for inductors and transformers, so in the first part of this paper, the magnetic properties of compositionally graded soft ferrites are reported for these applications.

Thermoelectric power generators have become a subject of great interest to meet the recent requirements for a reliable source of electric power. A number of materials have been tested for utilization in a thermoelectric generator. β-$FeSi_2$ is a promising material that useful for energy conversion in the temperature range of 200 to 900°C. This material is nonpoisonous, has good oxidation resistance, can be operated in air without any protection, and exhibits high electrical conductivity and large thermoelectric power.

The second part of this paper reports on the microstructure and thermoelectric properties of the $FeSi_x$ compound which have been investigated for the composition range x=1.90 to 2.20.

Bithumuth layer-structured ferro-electrics such as $Bi_4Ti_3O_{12}$ (BIT), $PbBi_2Nb_2O_9$ (PBN) and $Na_{0.5}Bi_{4.5}Ti_4O_{15}$ (NBT) are characterized [49,50] by (1) lower ε_r (100-200), (2) lower $\tan\delta$ (<0.005) and higher Curie temperature Tc (300-900°C). These properties are useful for piezoelectric and pyroelectric materials.

In the third part, the pyroelectric properties of $Bi_4Ti_3O_{12}$ based ceramics are investigated from the viewpoint of their application to pyro-sensor materials.

5.1.2 Magnetic Properties and Application of Multilayer Soft Ferrites

Ferrite samples were prepared by the conventional solid-state reaction of high-purity reagents, NiO, $MnCO_3$, ZnO and Fe_2O_3. The raw materials were mixed in an agate mortar with ethanol according to the chemical composition. The mixed powders were calcinated at 1000°C or 1100°C in air for 2 hours using an alumina crucible. The calcinated powder was inserted into a die and formed into pellets with a relatively lower pressure of 1000 kg/cm^2 as shown in Figure 4.47. Final samples were pressed under the pressure of 10 to 100 kg/cm^2. The pellets were sintered at 1340°C in air for 2 hours. They were cooled in-furnace to 1150°C at the rate of 5°C/min. and then quenched rapidly to prevent the re-oxidation of Mn^{2+} for samples containing Mn ions.

For the calcined powders, the phase identification was carried out by the powder X-ray diffraction method using Cu-K_α radiation. The sintered bodies were checked by SEM and EPMA analyses. The measurements of the magnetic properties were carried out using VSM (vibrating sample magnetometer).

The magnetic properties of multilayer soft ferrites such as $Ni_{1-x}Zn_xFe_2O_4$ and $Mn_{1-x}Zn_xFe_2O_4$ are described. Nickel ferrites ($NiFe_2O_4$) have a high Curie temperature (Tc≅600°C), but it decreases with the increase of x although the magnetization increases with x in the composition range of 0.4>x>0. In order to obtain Ni-Zn ferrites which have higher Tc and appropriate magnetization, multilayer $Ni_{1-x}Zn_xFe_2O_4$ with graded

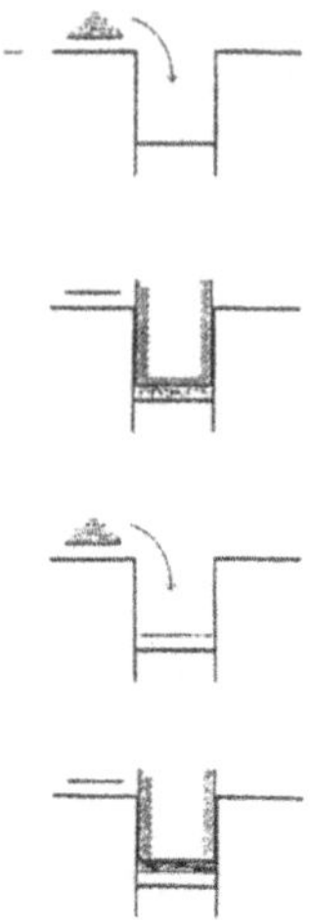

Figure 4.47. Pressing processes.

composition was prepared. Using four sheets of x=0, 0.1, 0.2 and 0.4 in the $Ni_{1-x}Zn_xFe_2O_4$ system, high Tc (≅600°C) and medium magnetization (σ=60 emu/g) ferrite was obtained as shown in Figure 4.48. Multilayer Ni-Zn/Mn-Zn ferrites were also prepared by using this method. Multilayer ferrites with average magnetization and high Tc may be useful for chip-size indicators and transformers.

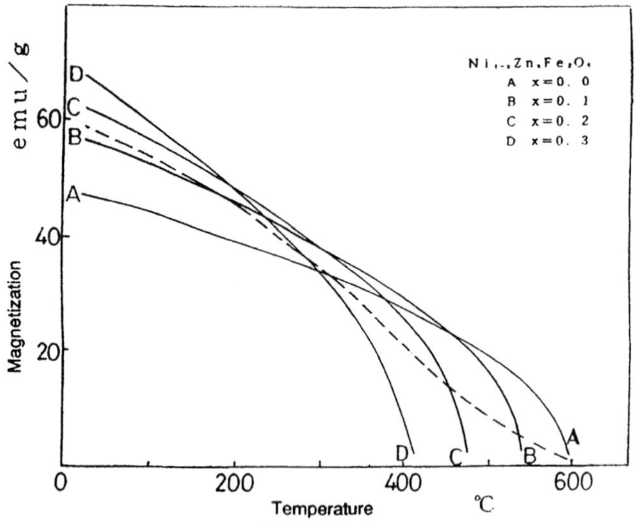

Figure 4.48. Temperature dependence of magnetization for $(Ni_{1-x}Zn_x)Fe_2O_4$ ferrites.

5.1.3 Effect of Microstructure on Thermoelectric Properties of $FeSi_x$ Compounds

The samples were prepared by the conventional powder metallurgy method described in the following. High purity (> 99.9%) metal powders such as Fe, Si, Mn and Co were used as starting materials. Mixtures of these powders were prepared in atomic percentages $FeSi_x$, $Fe_{1-y}Co_ySi_x$, and $Fe_{1-z}Mn_zSi_x$ ($2.1 \geq x \geq 1.90$, $0.05 \geq y$, $z \geq 0$).

The mechanical alloying (MA) was carried out in conventional stainless steel vials (1700 ml, 128 mm diameter) and balls (9.6 mm diameter). The spark plasma sintering (SPS) was performed in a carbon die with an 11 mm inner diameter. The SPS conditions were a vacuum of 0.1 Torr, load of about 40 MPa, and pulse direct current of 1300A at 3V, followed by heating at 1073K in air.

The crystal structure of the powders and sintered bodies was checked by X-ray analyses. Chemical compositions were analyzed by XPS. The electrical resistance and thermoelectric properties were measured by the 4-probe method.

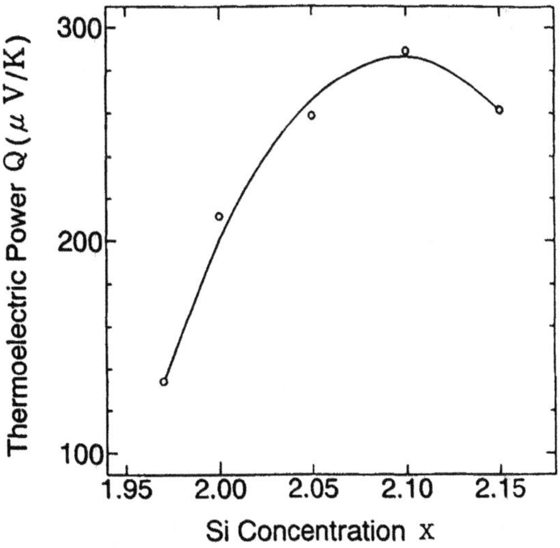

Figure 4.49. Si concentration dependence of thermoelectric power for $FeSi_x$(x=1.97 to 2.15) at 400K.

Figure 4.49 shows the thermoelectric power for the $FeSi_x$ compound as a function of the content of Si. In this figure the maximum value is obtained at x=2.1(Si/Fe). This may be due to the single phase of $FeSi_2$. It was confirmed by X-ray analyses that at lower x value, ε -phase appears at higher x value. Mn and Co doped $FeSi_2$ show p-type and n-type, respectively. Thermoelectric properties are also higher values at x=2.1 as listed in Table 4.5. For the MA+SPS processing, the temperature dependence of power factors are shown for $FeSi_2$ samples of spark plasma sintering at 1223K and annealing for 5 hours in air using various kinds of high-energy ball milling powders. Higher power factors were obtained at milling times of 250 ks. This may be due to the stability of the β-$FeSi_2$ phase. The thermoelectric properties of $FeSi_{2.1}$ are also higher than those of $FeSi_2$.

For higher conversion efficiency, higher figure of merit ($Z=(\alpha^2/k\rho)$ is desirable for thermoelectric devices. Here, α, k, and ρ are the thermoelectric power, thermal conductivity and resistivity, respectively. It has been reported [51] that functionally graded composition is useful for n-type PbTe. Thus, functionally graded compositions for $FeSi_x$ are also expected to improve thermoelectric properties. It may be expected that in Figures 4.50 and 4.51 the thermoelectric properties of functionally graded compositions of $FeSi_x$ are higher than those of $FeSi_2$.

Table 4.5. Thermoelectromotive force ΔE_o and effective maximum power W_{ell} at ΔT =600K, 800K for Mn doped and Co doped $FeSi_2$ sintered by SPS.

	ΔT=600K		ΔT=800K	
	ΔE_o (V)	W_{ef} (W·m/m²)	ΔE_o (V)	W_{ef} (W·m/m²)
$Fe_{0.95}Mn_{0.05}Si_{2.1}$	0.2	22	0.24	50
$Fe_{0.97}Co_{0.03}Si_{2.1}$	0.14	50	0.17	100

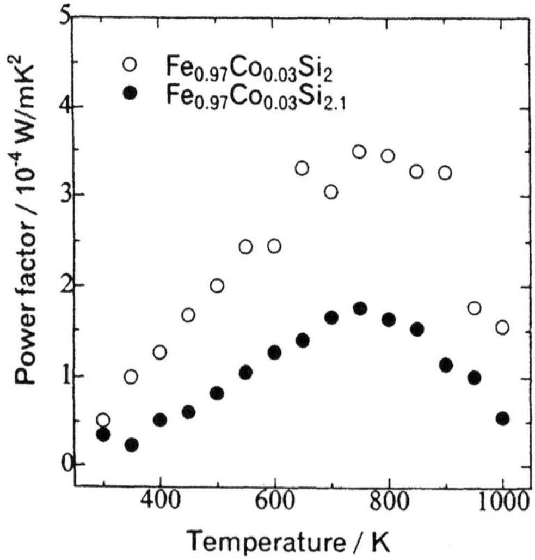

Figure 4.50. Temperature dependence of power factor for $Fe_{0.97}$ $Co_{0.03}Si_{2+x}$ samples of high energy ball milling for 500 ks and spark plasma sintering at 1223K and annealed for 5 hours in air.

5.1.4 Preparation and Pyroelectric Properties of Modified $Bi_4Ti_3O_{12}$ Ceramics

Reagent-grade oxide powders of Bi_2O_3,TiO_2,Nb_2O_5 and La_2O_3 with 99%+ purity were used as the starting materials. The powders of Nb_2O_5 doped $(Bi_{4-x}La_x)Ti_3O_{12}$ were uniformly mixed and ground in a ball mill. Ordinarily fired samples were prepared by the conventional sintering technique. The mixed powders were calcined at 800°C for 2 hours, then ground, pressed into disks and sintered at 1050°C for 2 hours. Grain-oriented samples were prepared by the hot-forged method at 1100°C for 2 hours under 200 kg/cm^2 pressure.

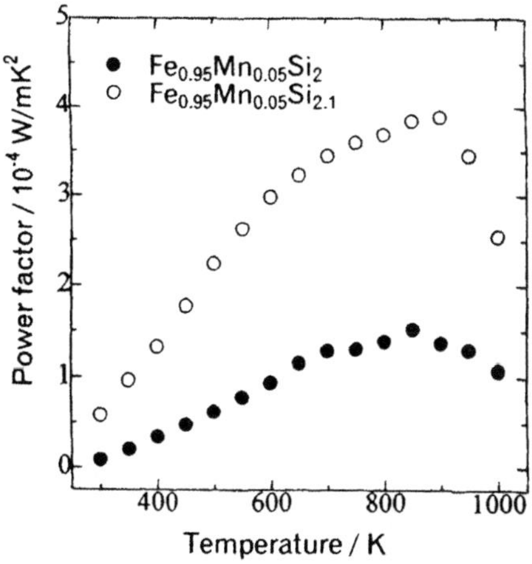

Figure 4.51. Temperature dependence of power factor for $Fe_{0.95}Mn_{0.05}Si_{2+x}$ samples of high energy ball milling for 500 ks and spark plasma sintering at 1223K and annealed for 5 hours in air.

The fired density was measured using the Archimedes method. The grain size was determined by the linear intercept method on the fired surface as observed with a scanning electron microscope (SEM). The crystallography was performed by X-ray diffraction (Monochromatized CuK α radiation). Capacitance was measured using same impedance analyzer at room temperature. The dielectric constant was calculated from the capacitance at a frequency of 1 kHz.

For the electrical and pyroelectric measurements, disk specimens 18 mm in diameter and 1.0 mm in thickness were formed from the sintered ceramics and were Au- or Ag-coated at 700°C for 5 min. The specimens were then immersed in silicone oil and poled in a 1.0 to 4.0 kV/mm DC field. The relatively lower field (1 to 2 kV/mm) was applied for non-Nb_2O_5 doped materials because of the lower resistance. However, the higher field (3 to 4 kV/mm) was applied for Nb_2O_5 doped samples with higher resistance. The electric field was applied at a temperature of 120°C for 10 min, and the specimens were then cooled to 30°C in the field.

The pyroelectric coefficient, p was measured by the modified Byer-Roundy technique over a temperature range from -20 to 200°C. The measurements were made during the heating cycles with a typical heating rate of 2 to 3°C/min. The coefficient, p was obtained from an averaged value of 20 to 25°C.

Generally, the pyroelectric coefficient p is described by the following equation: $p=(dP_s/dT)=(i_p/A)$ / (dT/dt). Here i_p is pyroelectric current defined by $i_p=A$ (dPs/dt) and A: dimension of electrode, *Ps:* spontaneous polarization, *T:* temperature and *t:* time.

To compare the properties of pyroelectric materials, figures of merit have been defined for the particular application of interest, and three values of which are expressed as follows [52],

$$F_i = p / C_v,\ F_v = p/(C_v \cdot \varepsilon_r),\ F_D = p/(C_v \cdot \sqrt{\varepsilon_r \tan\delta}\,) \quad (4.4)$$

where C_v is volume specific heat, ε_r is relative dielectric constant and $\tan\delta$ is dielectric loss. F_i and F_v are current and voltage responsivities, respectively and F_D is normalized detectivity which takes noise into account. Therefore, pyroelectric materials are required to have large p and low C_v, ε_r and $\tan\delta$ for high-performance detection of thermal radiation. Therefore, a larger p value in the wide temperature range close to room temperature is desirable for higher performance. Using the $(Bi_{4-x}La_x)Ti_3O_{12}$ +1.0 mol% Nb_2O_5 system, the functionally graded composition suitable for this purpose is shown in Figures 4.52 and 4.53.

5.1.5 Conclusions

For three functionally graded electronic materials, our results are as follows:

(1) For compositionally graded multilayer ferrites, magnetization has average value and Tc corresponds to the highest of those values.

(2) The thermoelectric properties of functionally graded compositions of $FeSi_x$ are improved compared with those of $FeSi_2$ in a wide temperature region.

Using the $(Bi_{4-x}La_x)Ti_3O_{12}$+1.0 mol% Nb_2O_5 system, samples with functionally graded composition may be suitable for pyro-sensor applications.

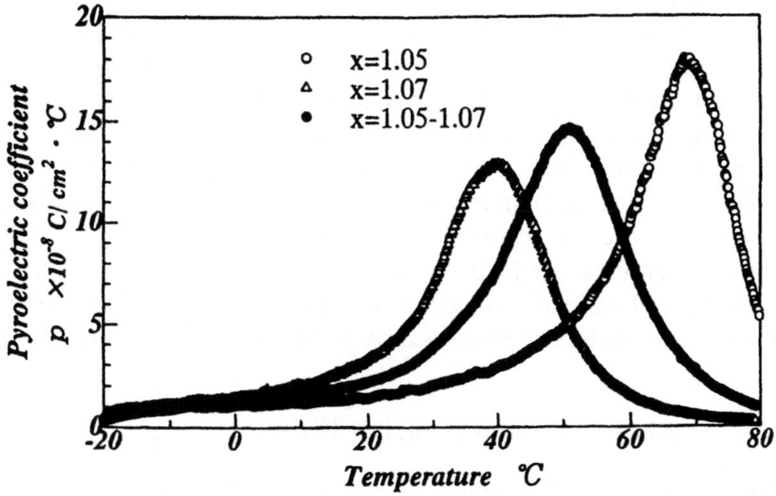

Figure 4.52. Temperature dependence of the pyroelectric coefficient of the $(Bi_{4-x}La_x)Ti_3O_{12}$+1mol% Nb_2O_5 system.

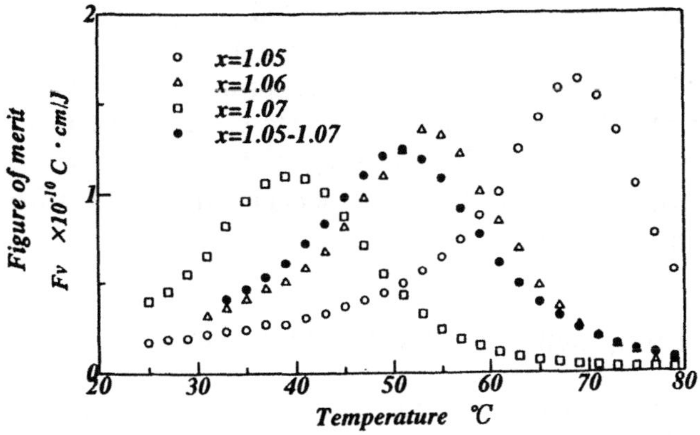

Figure 4.53. Temperature dependence of the figure of merit F_v in the $(Bi_{4-x}La)Ti_3O_{12}$+ 1mol%Nb_2O_5 System.

5.2 TiO_2-Based FGMs

5.2.1 An Explanation of Functionally Graded Materials

To produce a material that is effective for the relaxation of heat stress, it is necessary to distribute component materials in different proportions at the two faces. Over the past 15 or 20 years, development has been undertaken in Japan of this type of graded dual-functioned material, known as functionally graded materials, or FGMs. Thanks to the near-absence of interface effects,

FGMs offer a wide range of applications in many fields of engineering. In a typical FGM, the external surface consists of ceramic for resistivity to high temperatures, whereas the internal part is metallic to assure thermal conductivity and mechanical strength. Applications include: in electrical engineering, uses in radiators, sensors, magnetic insulators etc.; in mechanical engineering, uses in engines, machine tools, heat resistant materials etc.; and in building construction, uses in curtain walls, paved areas etc. The present report concerns itself with uses in electrical engineering, in particular as field relaxation materials, and reports on relative permittivity, dielectric loss tangent and breakdown voltage characteristics, as well as on these materials' effectiveness in the elimination of NOx gases.

5.2.2 Methods of Manufacturing FGMs

Korean kaolin and titanium oxide of uniform granular diameter are mixed in distilled water. The mixture is put into a cylinder and then vacuum filtered. When the first cake has been formed, fresh materials are put into the cylinder and thus successive layers are added [53]. After the final cake is formed, the whole FGM mass is subjected to applied pressure for 24 hours. The FGM is then dried naturally, and sintered to firmness in a reducing furnace. The titanium oxide used for the present test materials was 1st grade experimental TiO_2 with rutile and anatase type crystalline form. Korean kaolin is a clay primarily consisting of kaolin ore, having the chemical formula $Al_2Si_2O_5(OH)_4$ [54]. This manufacturing method has been granted patents in Japan and the USA [55].

5.2.3 Equipment Used for The Manufacture of FGMs

The apparatus was constructed in bronze and comprised four parts, an upper and a lower cylinder, a piston and two perforated plates. A schematic diagram of this apparatus is seen in Figure 4.54. The furnace used was manufactured by the firm Advantec. The furnace temperature can be controlled by computer, with an accuracy of ±1.5°C with respect to the set temperature. The internal dimensions are: height 20 cm, width 25 cm, depth 35 cm. After the sintering was completed, the furnace was left to return very gradually to normal temperature while a plentiful air circulation was assured. The furnace temperature was regulated in accordance with JIS R 8101-1959 [56].

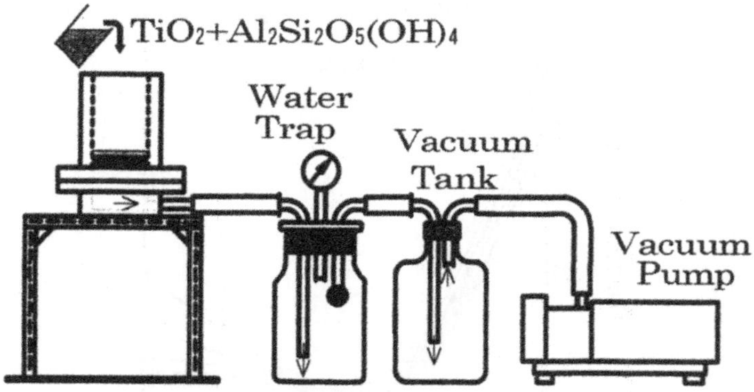

Figure 4.54. Apparatus for the manufacture of FGMs, by means of vacuum filtration.

5.2.4 Measurement of Relative Permittivity and Dielectric Loss Tangent

A Q-meter was used to measure the relative permittivity and dielectric loss tangent of the titanium-based FGMs sintered in an electric furnace. The sintering temperatures were 400°C, 600°C, 800°C and 1000°C, and measuring frequencies were 0.25 to 50 MHz. The values obtained for relative permittivity are shown in Figure 4.55, those for the dielectric loss tangent are shown in Figure 4.56. The TiO_2 used in the present experiment was mainly of the rutile type.

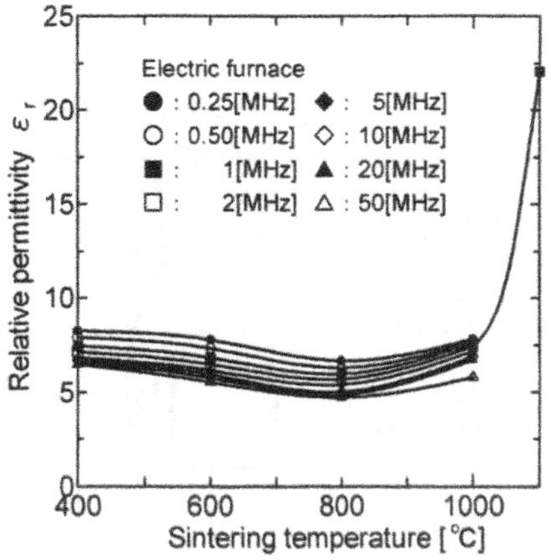

Figure 4.55. Relative permittivity as a function of sintering temperature for a TiO_2 FGM.

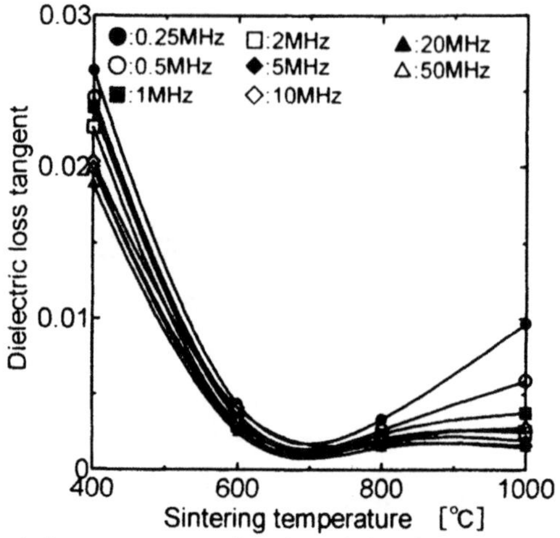

Figure 4.56. Dielectric loss tangent as a function of sintering temperature for a TiO_2 FGM.

5.2.5 Measurement of The Breakdown Voltage

The measurement was conducted in accordance with JIS C 2110 [57], using spherical electrodes of 12.5 mm diameter. The breakdown voltage in air was measured. The FGM piece was then inserted between the electrodes, voltage was applied, and the new breakdown point was measured. The piece was inserted in such a way that the TiO_2 face was in contact with the high voltage electrode, and the kaolin face with the low voltage electrode. Measurements were taken in this way for pieces produced at sintering temperatures of 400°C, 600°C, 800°C, 1000°C, and 1100°C. Subsequently, the thickness of each piece was also measured. The measurement circuit is shown in Figure 4.57 and the results are given in Figure 4.58.

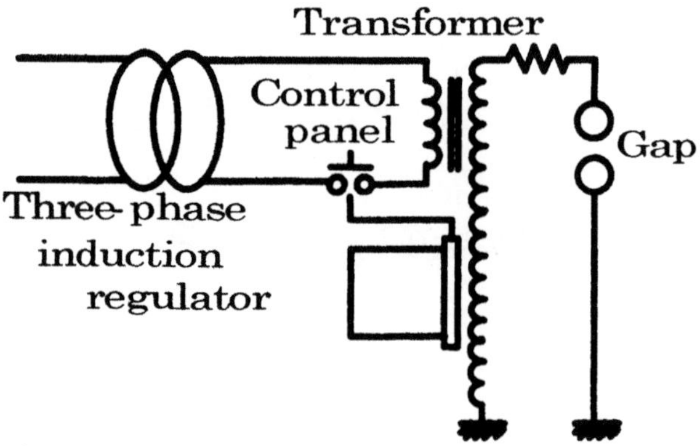

Figure 4.57. Circuit diagram of breakdown measurements for a TiO_2 FGM.

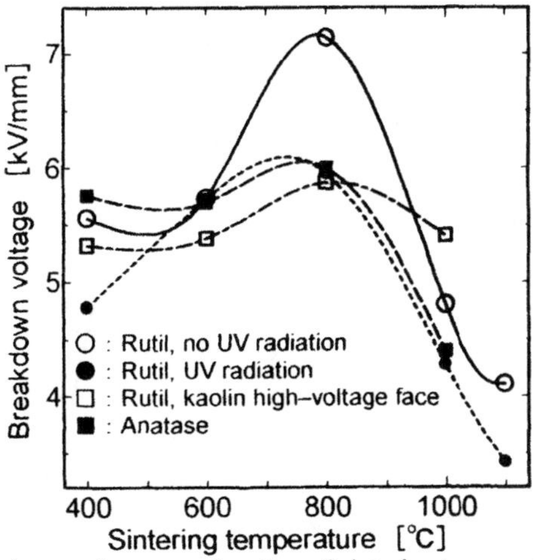

Figure 4.58. Breakdown voltage as a function of sintering temperature for a TiO_2 FGM.

5.2.6 Measurement of The NOx Removing Performance

For this test, five kinds of FGMs were used, manufactured at different sintering temperatures. After manufacture, the TiO_2 FGM pieces were placed in a pyrex container filled with two different kinds of standard gas (8.5 ppm NO_2; 10.1 ppm NO). They were then subjected to radiation from a 10 W lamp (wavelength 254-356 nm) for 2 hours, and finally immersed in 500 ml of distilled water for 1 hour. Flow injection analysis (FIA) was used to detect any release of nitrite ions. The apparatus for this test is shown schematically in Figure 4.59. The density of nitrite ions in the distilled water was measured using a UV-visible light detector. The results appear in Figure 4.60. The authors were using a single piece of test material of surface area $2.83\times10^{-3}m^2$ (diameter: 6 cm). The distance of the light source from the FGM was 20 cm.

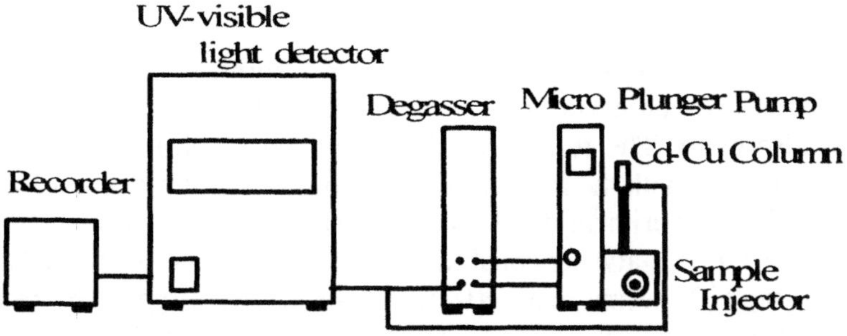

Figure 4.59. Schematic representation of apparatus for flow injection analysis.

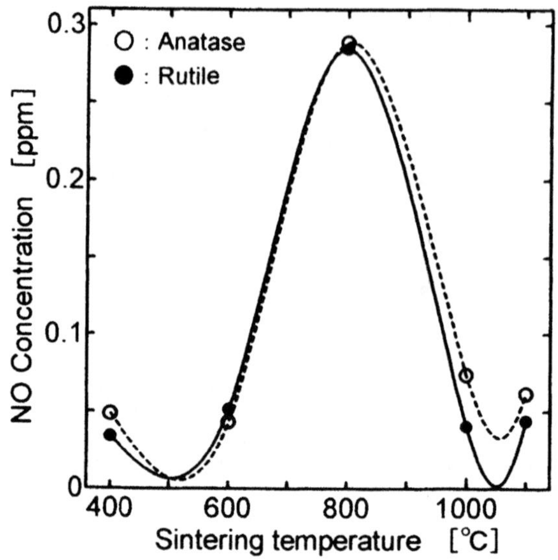

Figure 4.60. NOx adsorption as a function of sintering temperature for a TiO_2 FGM.

5.2.7 The Breakdown Voltage Experiment

The results reveal a rise in the breakdown voltage between the manufacturing sintering temperatures of 400°C and 800°C, with a peak at around 800°C. Further increases in the sintering temperature beyond this level lead to a decline in the breakdown voltage. The use of ultra-violet light led to a decrease in breakdown voltage for all the materials tested. A probable reason for this may be that the use of ultra-violet light causes a generation of electrons and protons on the surface of the FGM, which is what leads to the drop in the breakdown voltage.

5.2.8 The Photocatalysis Experiment

In general, NO is said to have a greater adsorption capacity than NO_2. But the authors' finding in this case was the contrary: the NO_2 showed the higher adsorption. The reason for this at present unclear. The result may partly have to do with the respective concentrations of the NO and NO_2. The findings of the experiment did, however, clearly show that this type of FGM possesses a photocatalysing capability. In the case of the photocatalyser, an anatase-type TiO_2 seems more effective than the rutile-type. The results of the experiment reveal no appreciable difference in NOx removing performance between the two crystalline types. From the point of view of the photocatalysing effect, the anatase form is usually said to offer the optimal characteristics. Of the rutile FGM materials tested in the present experiment, the results suggest that the materials sintered at 800°C have the better adsorption capacity for both NO and NO_2. From the point of view of the photocatalysing effect, the anatase form is usually said to offer the optimal characteristics. Of the rutile FGM materials tested in the present experiment, the results suggest that the materials sintered at 800°C have the better adsorption capacity for both NO and NO_2. For porosity characteristics, the results are similar.

5.2.9 Evidence of SEM Photographs

Looking at the SEM evidence, no very considerable differences are apparent. However, a greater number of pores can be seen in the material sintered at the lower temperature of 800℃. Part of the reason for this is the expansion of the kaolin grains around this temperature. In their future research, the authors intend to manufacture further types of FGM and to extend their investigations into their electrical and photocatalysing properties.

5.3 High-Performance Reflection Filters

5.3.1 Introduction

High-performance optical filters have a very important application in wavelength division multiplexing (WDM) for high-speed, large-capacity fiber-optic communications. Recently, the theoretical design and experimental manufacture of optical filters have attracted wide interest [58-68]. The material systems applied for optical filters mainly focus on oxides that have high transparency, low absorption and good reliability, such as the SiO_2-TiO_2, SiO_2-Ta_2O_2, SiO_2-ZrO_2, MgF_2-HfO_2 systems. Among them, the SiO_2-TiO_2 system has been investigated a lot because of the large differences in the refractive indices between TiO_2 and SiO_2.

Previously, Wang *et al.* successfully designed and fabricated SiO_2-TiO_2 system reflection filters with stepwise graded refractive index profiles using the concept of functionally graded materials (FGMs) [69,70]. They found that the graded refractive index profile is effective for eliminating the sidelobes in the transmittance spectrum. However, the results of structural analysis of the prepared SiO_2-TiO_2 multilayer films showed that columnar crystalline TiO_2 easily formed in the amorphous TiO_2 layers as the layer number increased. Such crystal inclusions have some degree of absorption and degrade the optical characteristics from the designed values. On the other hand, it has been reported that ZrO_2 film has a higher crystallization temperature (greater than 300°C to 400°C) than TiO_2 film (about 100°C to 300°C) [71-73]. SiO_2-ZrO_2 system films have also been investigated [74,75]. However, so far there has been no report on the SiO_2-ZrO_2 system multilayer film with graded refractive index profiles. In the present study, Wang *et al.* designed a new reflection filter of SiO_2-ZrO_2 system multilayer film with stepwise graded refractive index profiles.

5.3.2 Design Results and Discussion

For multilayer optical films with p+1 layers on a substrate, the transmittance spectra have been calculated by means of optical multilayer equations [76]. In general, the transmittance (T) and the reflectance (R) of light are represented by a simple equation, T=1-R, if the absorption in the system is ignored. Transmittance spectra with incident wavelengths at an incident angle of 0 were calculated using the refractive indices of SiO_2, ZrO_2 and SiO_2-ZrO_2 composite films. The data of the refractive indices shown in Table 4.4 were obtained from the experimental results on SiO_2, ZrO_2 and SiO_2-ZrO_2 composite films deposited by helicon plasma sputtering. The central wavelength used was 800 nm. The substrate is optical glass BK7 (n=1.52) with a thickness of 1 mm and so the reflectance on the interface of the substrate and air is placed into the calculated transmittance spectrum. The calculations for the designed multilayer films were performed using Kidger Optics FILM-2000 thin film design software.

Based on our previous design of a 31-layer SiO_2-TiO_2 system reflection filter with stepwise graded refractive index profiles, here Wang *et al.* use the configuration of this 31-layer multilayer stack as the starting configuration for the SiO_2-ZrO_2 system multilayer filter. This configuration of the multilayer filter can be written as follows:

$$\mathrm{BK7|LABCDECFBGAH(LH)^4LGAFBEDCBAL|Air},$$

where H, L, A, B, C, D, E, F and G are ZrO_2, SiO_2 layers and SiO_2-ZrO_2 composite layers with quarter-wave optical thickness as shown in Table 4.6, respectively. $(LH)^m$ shows L and H layers repeated m times. The profiles of LABCDE and EDCBAL in the configuration represent the multilayer structures with graded refractive indices. The relationship of the refractive index and the optical thickness of this filter is shown in Figure 4.61(a). The transmittance spectrum of this filter is shown in Figure 4.61(b). It was seen that there are high transmittances outside the stopband except for two small reflection peaks near the stopband. However, the reflectance of the stopband was less than 97.3%. To obtain the higher reflectance of the stopband, it is usually necessary to increase m of $(LH)^m$ because the ratio of the refractive

Table 4.6. Properties of SiO_2-ZrO_2 (CeO_2) composite films.

Film number	Film composition (mol:mol)	Refractive index (at 633 nm)
L	SiO_2	1.470
A	$0.965SiO_2$-$0.035ZrO_2$ (CeO_2)	1.535
B	$0.918SiO_2$-$0.082ZrO_2$ (CeO_2)	1.603
C	$0.823SiO_2$-$0.177ZrO_2$ (CeO_2)	1.666
D	$0.691SiO_2$-$0.309ZrO_2$ (CeO_2)	1.728
E	$0.456SiO_2$-$0.544ZrO_2$ (CeO_2)	1.802
F	$0.297SiO_2$-$0.703ZrO_2$ (CeO_2)	1.853
G	$0.180SiO_2$-$0.820ZrO_2$ (CeO_2)	1.917
H	ZrO_2 (CeO_2)	1.986

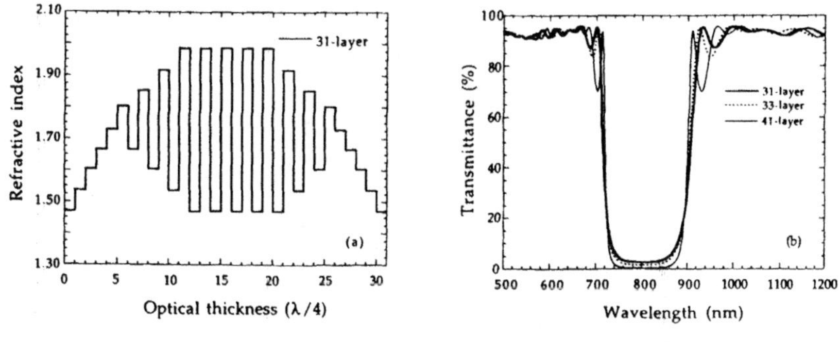

Figure 4.61. Relationship between the refractive index and optical thickness of a 31-layer filter with step wise graded refractive index profiles (a) and transmittance spectra of 31-, 33- and 41-layer filters with stepwise] graded refractive profiles (b).

indices of ZrO_2 and SiO_2 is less than that of TiO_2 and SiO_2. Thus, Figure 4.60(b) also gives the transmittance spectra of 33- and 41-layer filters with m=5 and 9. Figure 4.61(b) shows that the reflectance of the stopband increases up to greater than 99% in the range from 748 to 860 nm for the 41-layer filter. The cutoff characteristic became better, but the two peaks also became larger and closer to the stopband. Figure 4.62 shows the relationship between the refractive index and the optical thickness of the 41-layer filter. For the 41-layer filter, it is clear that the two small peaks can be suppressed by modulating the stepwise-graded refractive index profiles in the configuration of the filter. Sequentially, the approach of changing the optical thickness was also carried out. Figure 4.62 also shows the configurations of the two multilayer filters with different refractive index profiles. These are named 41-2D and 41-2D-2E with their formula as follows:

41-2D: BK7|LABC(2D)ECFBGAH$(LH)^4$LGAFBE(2D)CBAL|Air,

41-2D-2E:BK7| LABC(2D)(2E)CFBGAH$(LH)^4$LGAFB(2E)(2D) CBAL|Air,

where 2D and 2E represent the D and E layers with 2 times the optical thickness.

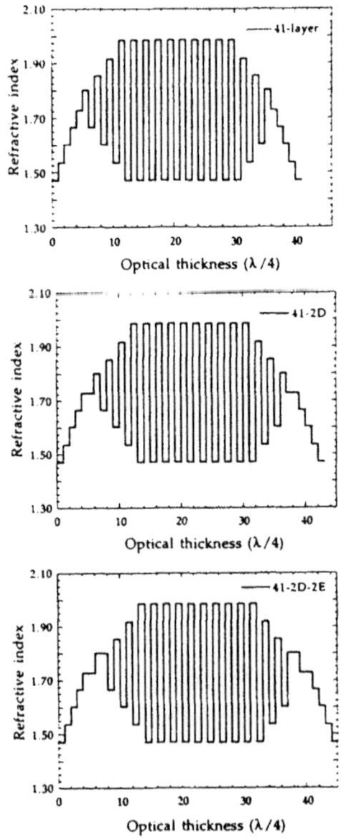

Figure 4.62. Relationships between the refractive index and optical thickness of three filters, 41-layer, 41-2D and 41-2D-2E.

Figure 4.63 shows the transmittance spectra of the 41-2D and 41-2D-2E filters. These filters exhibit similar optical characteristics compared to the 41-layer filter in Figure 4.61(b). The two peaks near the stopband could not be suppressed. Therefore, Wang *et al.* tried to design many configuration patterns for the multilayer filters and then calculated their transmittance spectra. As a result, a 42-layer filter with the following configuration was found to exhibit a good optical transmittance spectrum:

42-FGM:BK7|LABC(2D)(2E)CFBGAH(LH)^{4}LGAFB(2D)(2E)(2D)CBAL|Air.

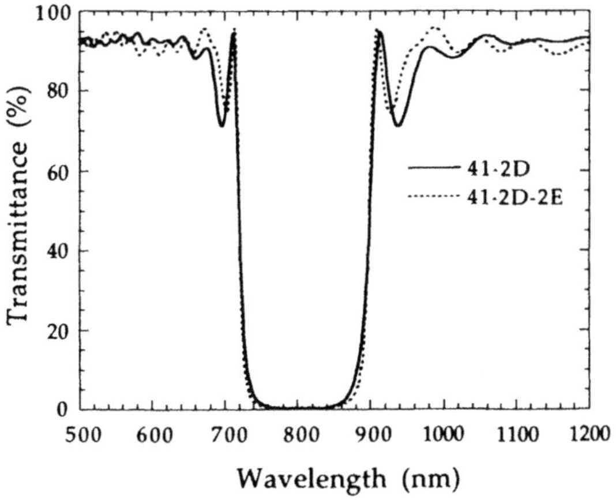

Figure 4.63. Transmittance spectra of 41-2D and 41-2D-2E filters.

Another 42-layer filter which has a similar structure also shows a good optical property and its structure is as follows:

42-S-FGM: BK7|LABC(2D)(2E)CFBGAH(LH)^{4}AGBFC(2E)(2D) CBAL|Air.

The configurations of these two filters are shown in Figure 4.64. From the above two formula and Figure 4.64, it is clear that there is a little difference at the modulating layers with alternating high- and low-refractive indices close to one side of the multilayer film surface.

The transmittance spectra of two of the three different filters are shown in Figure 4.65. It can be seen that two filters, 42-FGM and 42-S-FGM, have good optical characteristics. The stopband from 746 to 864 nm has high reflectance of greater than 99.0%, in which the reflectance at the reference wavelength of 800 nm is 99.7%. The full width at half maximum (FWHMs) of the stopband is 174 nm from 722 to 896 nm. In the transmission regions from 500 to 716 nm and from 906 to 2500 nm, high transmittances were obtained and the sidelobe was greatly suppressed by stepwise-graded refractive index profiles. From Figure 4.65, it can be seen that the 42-FGM has better optical characteristics than the 42-S-FGM. Therefore, it will be interesting to prepare a 42-FGM filter in advance. The deposition of a SiO_2-

ZrO_2 multilayer film by helicon plasma sputtering is in progress.

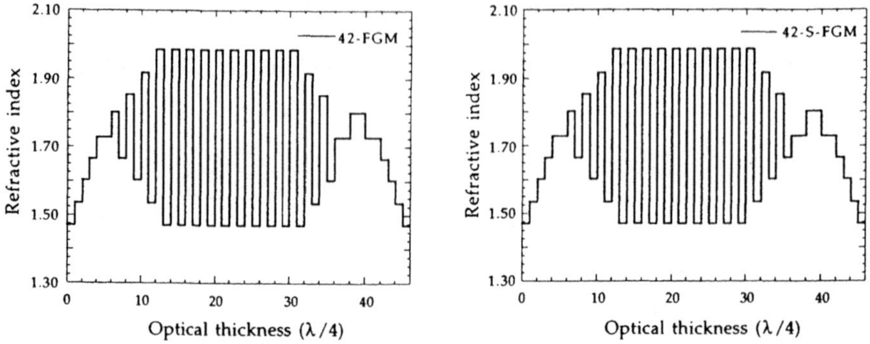

Figure 4.64. Relationships between the refractive index and optical thickness of 42-FGM and 42-S-FGM filters.

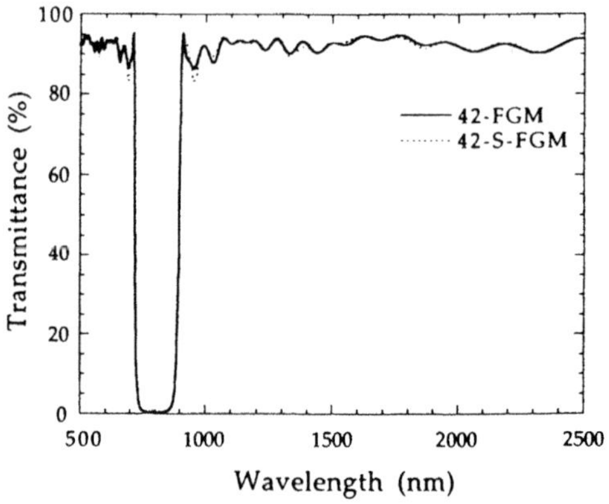

Figure 4.65. Transmittance spectra of 42-FGM and 42-S-FGM filters.

5.3.3 Conclusions

A 42-layer optical reflection filter with stepwise-graded refractive index profiles was designed. This is a new SiO_2-ZrO_2 multilayer reflection filter and has a configuration as follows:

$$LABC(2D)ECFBGA(HL)^{10}GAFB(2D)(2E)(2D)CBAL$$

where, A, B, C, D, E, F, G, H and L are the SiO_2-ZrO_2 composite films with different refractive indices and quarter-wave optical thickness. The transmittance spectrum of this filter exhibits sharp a cutoff characteristic and few sidelobes, and has high reflectances of greater than 99% in the stopband and high transmission in the pass fields.

5.4 Functionally Graded Thermoelectric Materials

5.4.1 Introduction

The target of Part II of the Japanese National Project on FGM is to maximize the efficiency of energy conversion materials utilizing the FGM concept. To improve the conversion efficiency of thermoelectric (TE) materials and devices, two different concepts of FGM configuration are being investigated. One is the application of graded dopant density in TE materials [77, 78]. The other is the application of multi-layered TE materials with graded potential barriers.

5.4.2 Optimum Design of TE Power Devices with Graded Dopant Density

5.4.2.1 Design of 3-Stage TE Power Generator Devices

Three-stage cascade and segment type TE power generator devices were designed, and the conversion efficiency of TE FGMs with the optimized graded dopant density calculated. Table 4.7 shows the graded dopant density and conversion efficiencies of power devices using optimized TE FGMs.

Table 4.7. Graded dopant density and conversion efficiency of optimized FGMs.

Material	Bi_2Te_3	PbTe	SiGe	Total
Temperature (K)	300~500	500~700	700~1100	300~1100
Graded Dopant Density (log 1/m³)	25.3~25.9	25.2~25.6	26.6~26.6	—
Efficiency of Cascade Module (%)	7.68	5.24	6.29	18.02
Efficiency of Segment Module (%)	7.68	5.24	6.29	15.42
Efficiency of SiGe Single Module (%)	—	—	—	12.32

The efficiency of the cascade type module is higher than the segment type module because Peltier heat generation between different TE materials is much higher in the segment module which causes loss of energy conversion. Figure 4.66 shows the optimum designs for 3-stage cascade and segment type TE generator devices using FGMs with graded dopant density.

5.4.2.2 Thermal Stress Analyses

Thermal stress analyses were performed for 3-stage cascade and segment type TE devices by FEM. Analysis models and the calculated maximum principal stress distributions are shown in Figures 4.67 and 4.68.

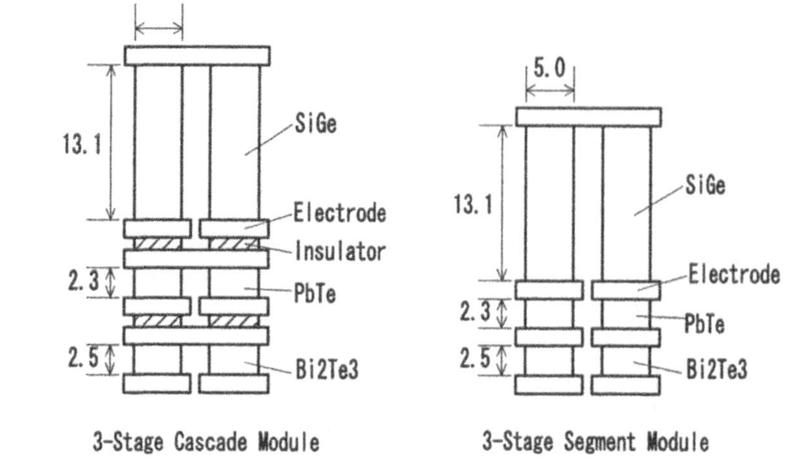

Figure 4.66. 3-stage cascade and segment type TE generator devices.

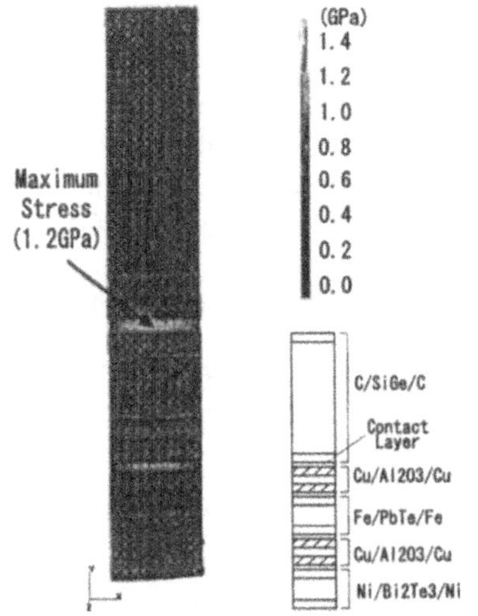

Figure 4.67. Maximum principal stress distribution in 3-stage cascade device.

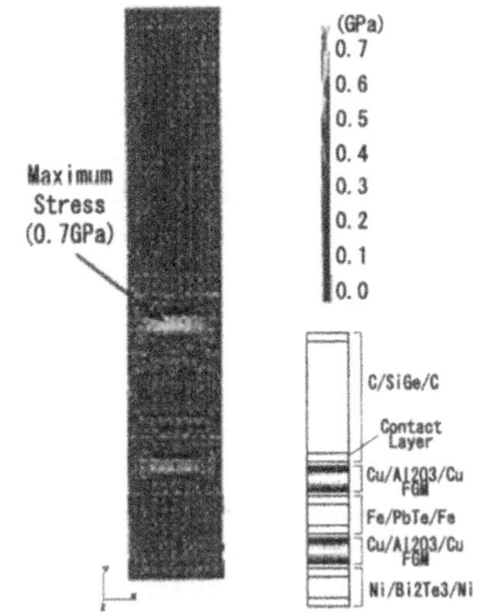

Figure 4.68. Maximum principal stress distribution in 3-stage cascade device with FGM insulation layers.

The cascade type model was constructed with a ceramic electrical insulator layer, copper electrode, contact layer, Ni/Bi_2Te_2/Ni, Fe/PbTe/Fe and C/SiGe/C TE materials (Figure 4.67). The thermal stress is concentrated at the interface between the electrode and ceramic insulator, and the maximum stress value is over 1.2 GPa. It was estimated that the ceramic insulator will fail due to the thermal stress. The segment type model was constructed with a copper electrode, contact layer, $Ni/Bi_2Te_3/Ni$, Fe/PbTe/Fe and C/SiGe/C TE materials, and the maximum stress value is about 260 MPa. Because the cascade type module has more efficient TE performance than the segment module, and the thermal stress is much severer than in the segment module, the FGM insulation layers are designed for thermal stress relaxation. Figure 4.68 shows the stress distribution in the cascade module with FGM insulation layers, and the maximum principal stress is significantly decreased to 700 MPa.

5.4.3 Multilayered Thermoelectric Materials

In addition to the graded composition of thermoelectric materials, it may be possible to improve their efficiency by using potential barriers since contributions from the majority carriers with low energy degrade the efficiency [79]. Hirano *et al.* briefly report on the method and the results for thermoelectric multilayers.

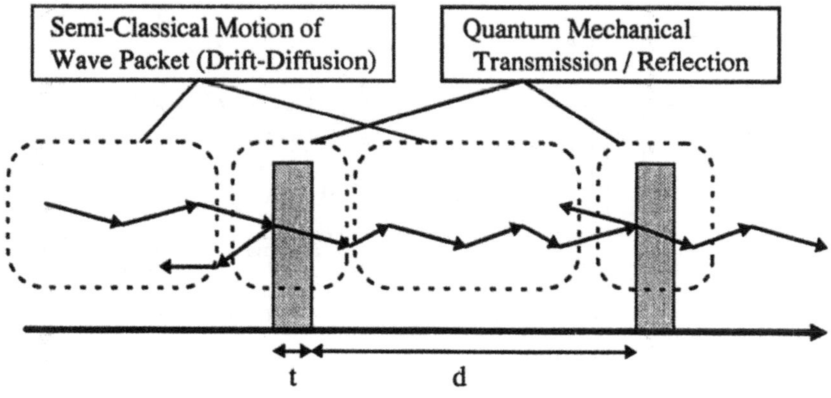

Figure 4.69. Kinetic formulation of transport.

5.4.4 Carrier Transport Across Multilayered Structures

First Hirano *et al.* will summarize the method of calculating the transport coefficient. The details of the calculation will be published elsewhere. The spatial region has been divided into two parts according to the characteristics of the carrier transport which depends on the length of the scale (Figure 4.69) [80]. For simplicity, they neglect the charge accumulation (depletion) and heat generation (absorption) at the barrier layer [81, 82]. Under these approximations, the transport properties were calculated for a two-band parabolic model (Figure 4.70). A moderate increase of the figure of merit had been expected (Figure 4.71).

5.4.5 Multilayer Fabrication Processes

One of the methods for fabricating the sample with the multilayered structure or graded alloy composition is electrodeposition [83, 84]. By changing the voltage of the working electrode one can easily change the alloy composition of the deposit. Since, for the multilayered sample, there exists a finite deviation of the alloy concentration at the boundary between layers, Hirano *et al.* have used process simulation to clarify the experimental condition for a given multilayered structure by estimating the thickness of the deviated region.

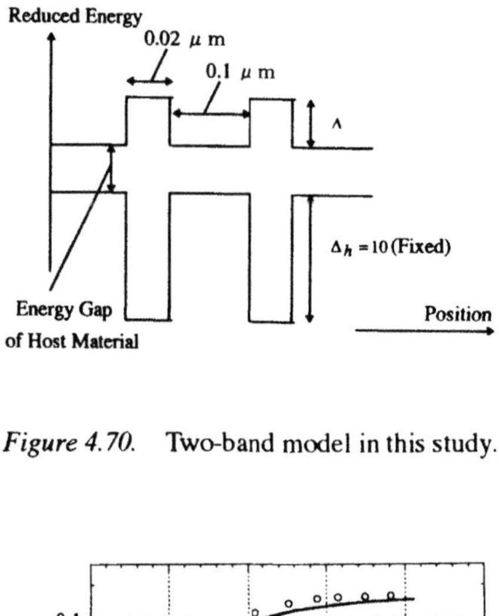

Figure 4.70. Two-band model in this study.

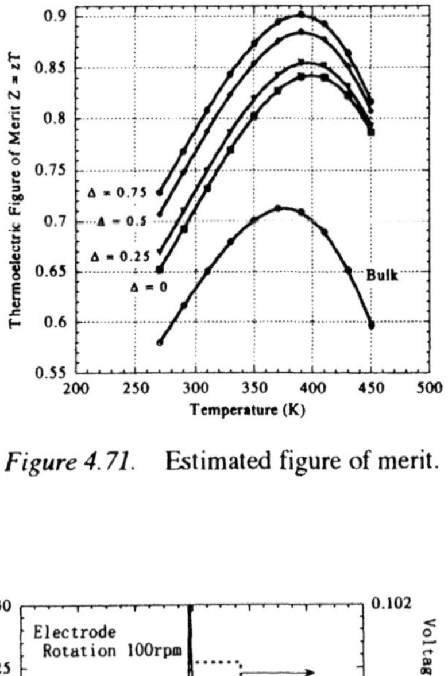

Figure 4.71. Estimated figure of merit.

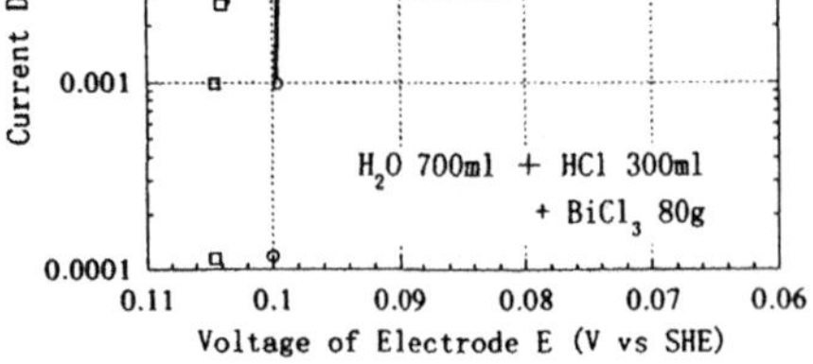

Figure 4.72. I-V Characteristics for BiSb.

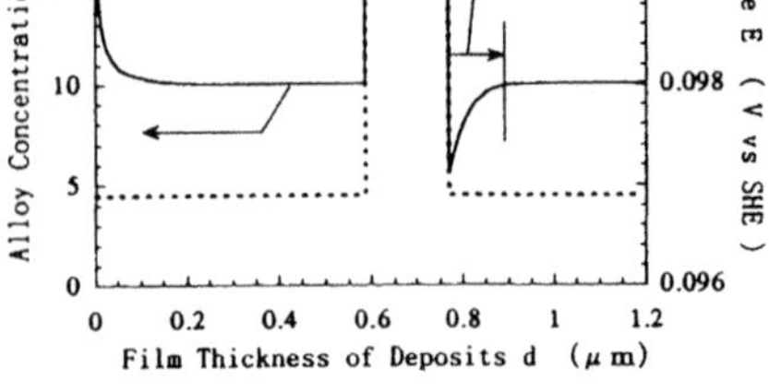

Figure 4.73. Fabrication of multilayer.

5.4.6 Simulation Example

The procedure of the process simulation is essentially the same as the one in the references[83]. The parameters in the simulation were determined to reproduce the observed current-voltage characteristics for BiSb electro-deposition (Figure 4.72). Figure 4.73 corresponds to the compositional profile for the experimental conditions given in Figure 4.72. Hirano *et al.* can see that the thickness of the deviated region is about 0.1 micrometer, thus multilayers with layer thicknesses comparable to 0.1 micrometer cannot be fabricated under these experimental conditions. Hirano *et al.* should thus change the experimental conditions to fabricate the multilayer with a layer thickness less than 0.1 micrometer.

6. TRIBOLOGICAL PROPERTIES

6.1 Tribological Approach

6.1.1 Introduction

Functionally graded materials (FGMs) have two chief advantages of over coatings. Firstly a gradual change of composition and structure avoids the problems of local shear stresses and adhesion; secondly, the technology allows to obtain compressive stresses in the surface layer that decrease fracture and fatigue failures. Knowledge of the different wear modes and the material properties that influence them permits an effective design of materials for optimal performance. A survey of the literature [85] shows few in-depth investigations of such materials other than the claim that they improve wear resistance. Fischer *et al.* review the wear mechanisms that operate in hard materials and in coatings in order to analyze the conditions under which increased wear resistance can be achieved by a continuous variation of composition. A rapid look at the wear rates reveals that sliding wear is about 10,000 times slower than abrasion. Thus, cutting, plastic deformation and fracture are the phenomena characterizing abrasion and fatigue phenomena play a major role in sliding wear. More information about wear resistance can be gained by an examination of the operating wear mechanisms.

6.1.2 Abrasive Wear

All mechanical wear is a response of the material to the stresses caused by contact. Abrasive wear results from the penetration of hard particles into the surface. The depth of penetration depends on the hardness of the material, thus abrasive wear is described as

$$V = k \frac{P}{H} s \qquad (4.5)$$

where V is the volume removed, P the total load of the abrasive on the surface, H the hardness of the abraded material and s the sliding distance. The equation predicts that abrasion resistance is proportional to hardness. *k* is known as the Preston coefficient; it contains the effect of other material properties, especially ductility.

Hard materials provide abrasion resistance only when they are harder than the abrasive. Failure of the coating then is caused by fracture or detachment from the substrate, both of which are alleviated by the graded composition and the compressive surface stresses of FGMs. When the abrasive is harder than the surface, its groves a brittle material by fracture and the Preston coefficient *k* is much higher than in metals[86]. In this case, a ductile material provides better protection than a brittle, hard, ceramic.

6.1.3 Sliding Wear

Sliding wear is the response to the contact stresses caused by load and friction of two non-penetrating surfaces. An understanding of the wear of hard materials is facilitated by an examination of these contact stresses.

6.1.3.1 Contact Stresses

Surfaces contact each other at asperities. In metals, the asperities deform plastically under the load; thus the local compressive stress under the asperities is equal to the hardness and independent of the applied load [87]. Wear is then proportional to the real area of contact A_R = P/H and is quite well described by Equation (4.5) where *k* is now called the wear coefficient and is generally 10,000 times smaller than in abrasion.

In ceramics, asperities deform elastically. In first approximation, an increase of the load increases the density of contacting asperities and the real contact area[88], as a corollary, the local surface stress at each contacting asperity depends little on the load; it is much larger than stresses calculated from the macroscopic geometry [89].

At shallow depths, (in the micrometer range) the stress underneath each asperity diminishes in intensity and spreads out laterally as the depth increases. It can be described, on a microscopic scale, by the formalism of Hamilton and Goodman. As the depth increases to a value in the range of the asperity distances on the surface, the stresses caused by the individual asperities overlap and the stress at any point is the sum of stresses caused by neighboring asperities [89]. Since this distance decreases with increasing load, the sub-surface stresses increase with the load. They are still larger than the stresses calculated form the macroscopic geometry. Gradually, these stresses decrease and approach the macroscopic values described in the next paragraph.

At the other extreme, namely a depth much larger than the distance between neighboring asperities, the contact mechanics is adequately approximated by the designed contact, neglecting surface roughness. For the discussions of wear phenomena to follow, Fischer *et al.* call these "macroscopic stresses".

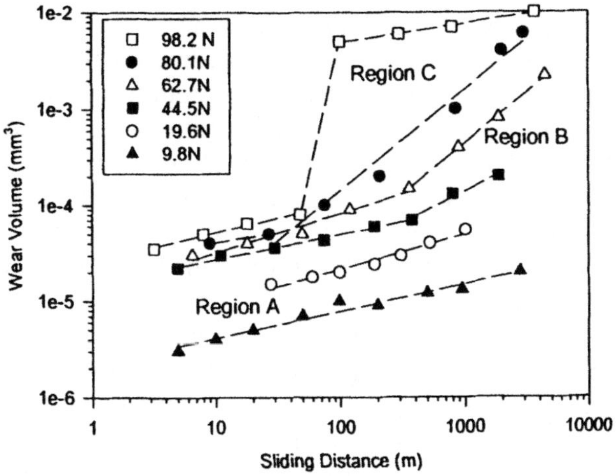

Figure 4.74. Sliding wear of Al_2O_3.

6.1.3.2 *Mechanical Wear of Brittle Materials*

The wear of brittle materials seldom obeys Equation (4.5). Its complexity is evident in the case of alumina [90-92]. Figure 4.74 shows the evolution of wear as a function of sliding distance and contact load in the case of unlubricated sliding of a polycrystalline alumina against itself. Fischer *et al.* distinguish three different behaviors that can be rationalized in terms of the stress distributions described under Section 6.1.3.1.

At low loads ($\leq$ 19.6 N) the amount of material removed is proportional to the load. The wear rate is very small, at 3000 m sliding distance and 9.8 N load, they compute a wear rate of 7.10^{-10} mm^3/Nm which is about 100,000 lower that in most metals! Microscopy of the surfaces reveals generally smooth surfaces with some grooves and fracture on a very small scale, much smaller than the grains.

When the load is increased (up to 80 N in Figure 4.73), the wear mode just described proceeds for a short distance, then changes to a mode where the volume of material removed increases with the 5th power of the load (doubling the load from 44.5 to 98.2 N increases the wear rate 30 fold, Regime B on Figure 4.74). The delayed onset of this regime suggests a fatigue phenomenon. Removal of the debris uncovers the loss of entire grains by intergranular fracture. A Vickers indentation in the wear scar produces tilting of some grains, showing that rubbing has weakened the grain boundaries. Obviously, the wear proceeds by grain boundary fatigue. This phenomenon occurs in the shallow layer below the surface described in Section 6.1.3.1: the stresses a few micrometers depth are lower than at the surface, they increase with the load and are sufficient to cause fatigue of the material. The rapid increase of the fatigue-cracking rate with stress suffices to explain the large effect of the load on the wear.

When the load becomes large, the macroscopic stresses are large enough to cause propagation of a macroscopic crack from the surface. Such cracks initiate at the rear of the contacting slider (pin) where maximum tensile stresses are largest. This causes the removal of large chunks of material and constitutes "catastrophic wear" at high loads [91, 92]. This is regime C in Figure 4.73. This wear regime depends of the strength of the grain boundaries. If these latter are relatively weak, the large fatigue cracks do not occur, but are replaced by diffuse intergranular fracture (crumbling) below the surface [93] and catastrophic wear is reduced.

For ceramics, the fracture toughness is a more reliable predictor of wear resistance than hardness [94]. In zirconia, the wear resistance is proportional to $(K_{1C})^4$. The compressive surface stresses that can be achieved in FGM's can be expected to decrease the incidence of wear by fracture, especially the large cracks of catastrophic wear.

6.1.3.3 Tribochemical Wear

Chemical phenomena influence the wear of ceramics in various ways, depending on the material, the environment and the mechanical conditions of rubbing. This influence can consist in chemically induced cracking that increases wear rates [95], in modifications of surface composition and topology that decrease wear [96], and in a purely chemical form of wear by dissolution in the liquid environment.

In zirconia, humidity increases the wear rate by at least one order of magnitude. No reaction products are observed, but intergranular fracture is increased by the phenomenon of adsorption embrittlement [88]. Usually, this phenomenon increases the rate of intergranular fatigue fracture, with the removal of plates that expose the grains.

Silicon nitride and silicon carbide undergo a tribochemical reaction with water molecules. This reaction is stimulated by friction and occurs at contacting asperities, which are removed, leaving a smoother surface. The result is a reduction of the local stresses responsible for the mechanical wear and a large decrease in the wear rate. When sliding occurs in humid air, the reaction product is a hydrated silica that covers the surface and further reduces local contact stresses. When sliding occurs in water or aqueous solutions, the material of the contacting asperities is dissolved and no wear particles are obtained. This phenomenon is used as a polishing method that produces surfaces with a roughness Ra = 0.5 nm.

In the case of alumina, humidity produces a surface hydride that reduces friction [91, 97]. This hydride is removed by friction to form the rolls described above. Very modest embrittlement by humidity has been observed for this material [90].

6.1.4 Effects of Processing

The processing of materials often introduces defects into a material that prevent us from obtaining the advantages sought. In thermal sprayed WC/Co coatings, for instance, decarburization, porosity and limited inter-splat adhesion are weaknesses that overpower the advantages one expects form refinements such as graded compositions or nanoscale microstructure. Fischer *et al.* have measured the wear resistance of two WC/Co coatings, one of which is straight WC-17% Co and the other functionally graded (Figure 4.75). The functionally graded material shows lower hardness and a wear resistance that is commensurately lower than that of the conventional material. Both materials wear by the removal of splats. The sectioning of the material to observe the microstructure in the cross-section of the coating caused a fraying of the material at the edge. Observation of this fraying revealed a material removal by breaking-off of individual splats, just as in the sliding wear. Other advantages of the functionally graded material may, of course, be a higher resistance to debonding of the coating, such debonding, however, is rarely a problem in WC/Co coatings.

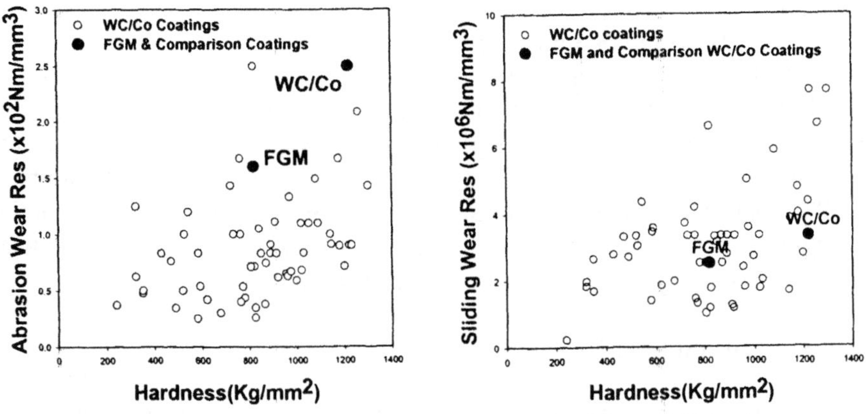

Figure 4.75. Abrasion and sliding wear resistance of WC/Co and FGM coatings.

Fischer *et al.* conclude that graded compositions will obtain their intended benefits only when the processing of the surface does not introduce material weaknesses of its own.

6.2 Tribological Properties and Damage by Pin-on-block Friction

6.2.1 Introduction

A new ZrO_2-stainless steel functionally graded material (FGM) is expected to increase the running temperature of an engine for higher performance. Although the production technology, residual stress, heat shock resistance and other characteristics of this FGM have been widely studied, there are only a few papers on its tribological evaluations.

In a previous study [98], cracks and chipping were generated in the ZrO_2 layer. But detailed conditions of the damage generation are not clear. As the generation of fatal damage such as cracks and chipping must never be permitted in practical use, the purpose of the present study focused mainly on a detailed investigation of these phenomena.

6.2.2 Preparation and Testing

ZrO_2 (3% Y_2O_3 PSZ)-stainless steel (SUS410L) FGMs with 11 layers made from granulated 50 μm PSZ and 9 and 60 μm SUS powders were supplied by Izumi Technology Co., Ltd. For the friction tests, the FGMs were cut into 3-mm thick $10^W \times 15^L \times 6^T$ mm PSZ and 90 to 70% PSZ-SUS layers. The surfaces of the PSZ layers were polished with a 800 mesh diamond wheel. Some specimens were treated by a special process to improve resistance to friction damage. The codes for the specimens used are as follows:

9U: 9 μm SUS, untreated
60U: 60 μm SUS, untreated
60T: 60 μm SUS, treated

The friction tests were carried out between the PSZ layer and a PSZ pin with a 4 mm top radius with a pin-on-block reciprocal friction tester. The tests were done at a load of 10N, temperatures from room temperature to 800°C friction speed of 1.2 m/min, friction time of 120 min, total friction length of 144 m in air. The friction coefficients and specific wear rates were measured and the damage to the PSZ was observed.

6.2.3 Tribological Properties

The friction coefficients and specific wear rates of the PSZ layer obtained by 120 min friction are shown in Figures 4.76 and 4.77. However as described later, after the friction tests, cracks and breakdown were found in the PSZ layer at and above 550°C in the 9U and 60U specimens and at 750°C in the 60T specimens.

Although the friction coefficients were fairly scattered, there was a tendency to decrease with the friction temperature. The friction coefficient of 9U using a fine SUS powder was relatively higher than the coarse 60U. No remarkable difference was observed between the treated and untreated specimens.

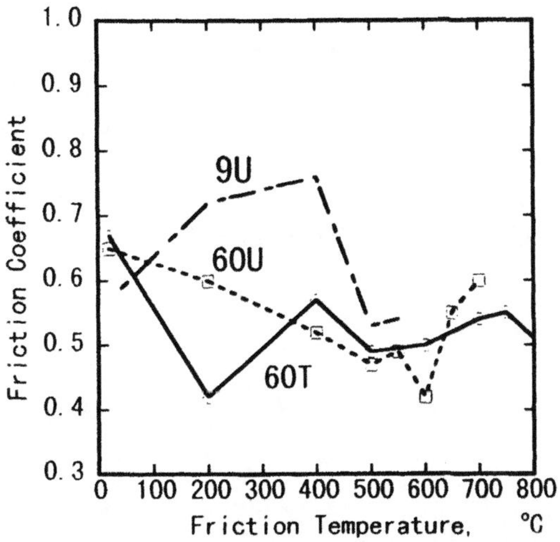

Figure 4.76. Friction coefficient between PSZ layers and PSZ pin.

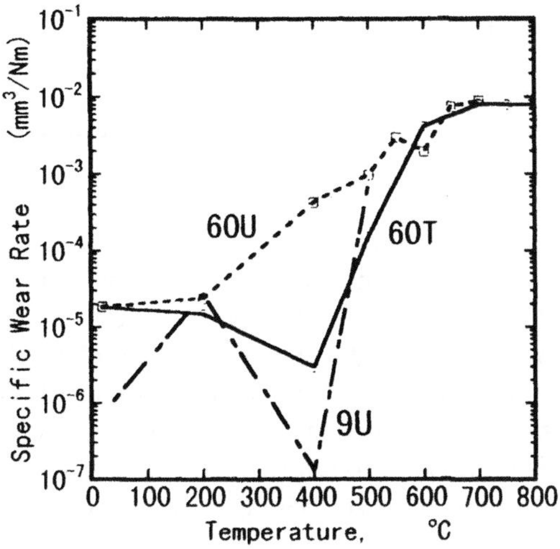

Figure 4.77. Specific wear rate of PSZ layers.

The specific wear rate at low temperatures was mostly in the order of 10^{-5} to 10^{-7} mm^3/Nm, but it remarkably increased to an order of $10^{-3}mm/Nm$ at high temperatures over 500°C. The fine 9U showed smaller specific wear rate than the coarse 60U. The effect of treatment on the specific wear rate was not clear as the difference at 400 and 500°C did not confirm whether it was essential or not. The specific wear rate of the PSZ pin was generally one order smaller than the PSZ layer throughout all temperatures.

6.2.4 Damage of PSZ Layer by Friction

The surface of the PSZ layers and interface of the PSZ and 90% PSZ-SUS layers after friction are shown in Figures 4.78 and 4.79. Damage of the PSZ layer at each temperature is summarized in Table 4.8. No cracks were observed on the surface of the PSZ layer after friction below 500°C in the 9U and 60U specimens and at 700°C in the 60T specimen. As shown in Figure 4.78, 9U broke down at 550°C. Vertical cracks were gene-rated by friction over 550°C in 60U and 750°C in 60T.

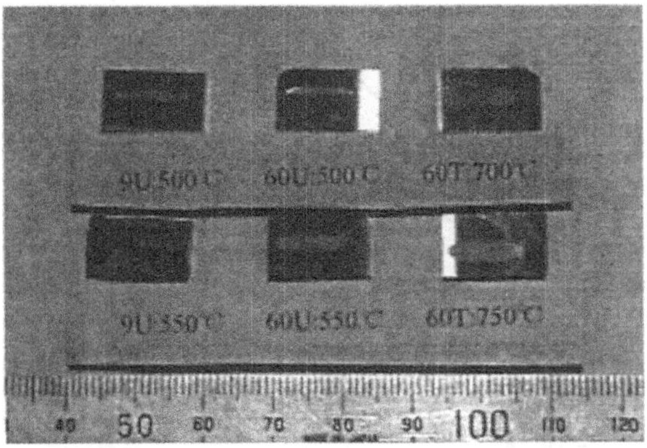

Figure 4.78. Surface of PSZ layers after friction.

Figure 4.79(a) shows the interface of the PSZ and 90% PSZ layers without cracks or pores. The normal interface remained after friction below 400°C in U, 500°C in 60U and 700°C in 60T, respectively. Hairline cracks and micropores were generated near the interface of the PSZ and 90% PSZ layers in 9U after friction at 500°C as shown in Figure 4.79(b).

These hairline cracks are considered to be the origin of the vertical cracks. The vertical cracks through the PSZ layer shown in Figures 4.79(c) were generated by friction over 550°C in 60U and 750°C in 60T. Some cracks penetrated the PSZ-SUS layers.

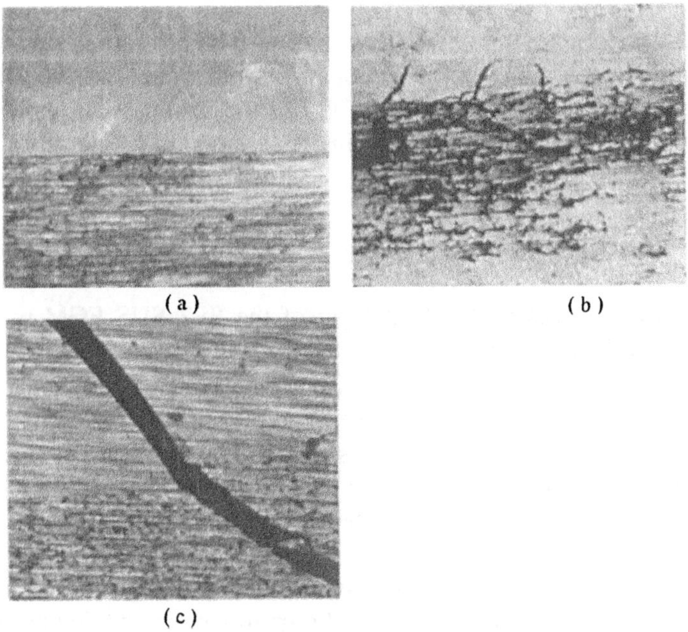

Figure 4.79. Interface of PSZ and 90% PSZ-SUS layers after friction.

Table 4.8. Damage of PSZ layer generated by friction.

/°C	9U	60U	60T
RT	No	No	No
200	No	No	No
400	No	No	No
500	B crack, B pore	No	No
550	V crack, Breakdown	V crack	
600		Small chipping	No
650		Medium chipping	No
700		Large chipping	No
750		Breakdown	V crack
800			V Crack

B crack: boundary crack, B pore: pores near boundary
V crack: vertical crack through PSZ layer

As can be seen in Table 4.8, the fine grain 9U showed the lowest crack resistance. The vertical cracks generated at 550°C in 60U led to chipping at 600°C and the size of the chipping increased as the friction temperature rose.

Then the 60U specimen broke down at 750°C. The higher crack resistance of 60U than 9U is considered to be due to the larger grain size of the SUS powder.

The treated coarse grain 60T showed the highest crack resistance. The maximum friction temperature without crack generation was elevated 200°C by pretreatment from 500°C to 700°C. As the results clearly show, special pretreatment is a promising way to improve the high-temperature performance of a PSZ-SUS FGM.

6.2.5 Conclusions

The friction tests on the PSZ layer of the PSZ-SUS FGM using a PSZ pin revealed the following results:

(1) When the friction temperature rises, the friction coefficient decreases, while the specific wear rate increases, especially over 500°C.

(2) The friction coefficient decreases, while the specific wear rate increases at larger grain sizes of the SUS powder.

(3) A coarse SUS grain exhibits a better resistance against crack generation and breakdown.

(4) The maximum friction temperature without crack generation is elevated 200°C by special pretreatment. Pretreatment is the important key to improving the high-temperature performance of a PSZ-SUS FGM.

6.3 Wear Resistance

6.3.1 Introduction

The wear resistance of conventional tribomaterials is generally poor in low-viscosity fluids because the fluid cannot form a lubrication film on the sliding surfaces. Si_3N_4-TiN-Ti functionally graded material (FGM) is designed for a tribomaterial to be used in low-viscosity fluids. The base material is Ti because Ti exhibits good toughness and corrosion resistance. The surface of the material is a Si_3N_4 rich layer because Si_3N_4 has good wear resistance[99-101]. TiN is a sintering additive. To find the optimum surface composition, composites with various mixtures were produced by spark plasma sintering (SPS). Sliding tests were conducted under various conditions in which the Si_3N_4-TiN-Ti composites were slid against a Si_3N_4 ceramic ball purified with n-hexane, in distilled water, anhydrous ethanol and in a dry condition. The worn surfaces were examined using a scanning electron microscopy (SEM), an electron probe microanalyzer (EPMA) and energy dispersive spectroscopy (EDS).

6.3.2 Preparation of the Si_3N_4-TiN-Ti Composite

α-Si_3N_4 powder, TiN powder and Ti powder were mixed in a planetary micro-pulverizer (Fritsch P-7). The mixed powders were sintered using an SPS apparatus (Sumitomo Coal Mining Co., Ltd., Model SPS-511S) at a pressure of 64 MPa and at in-disk temperatures of 1300°C to 1500°C (diameter: 20 mm, thickness: 4 mm). The Si_3N_4-TiN-Ti composite containing more than 40 wt. % of Si_3N_4 did not solidify under these conditions. For comparison, a Si_3N_4 ceramic disk (Si_3N_4-Al_2O_3-Y_2O_3: 90-5-5 wt.%) was produced using the same procedure. One of the disk surfaces was polished using a number 1000 grade diamond grinding wheel. The composition, density and hardness (Hv) of the sintered materials are shown in Table 4.9.

Table 4.9. Composition, density and hardness (Hv) of the sintered materials.

Material	Composition (wt. %)			Density (g/cm^3)	Hardness, Hv (GPa)
	Si_3N_4	TiN	Ti		
0% Si_3N_4	0	10	90	4.5	6.0
1% Si_3N_4	1	10	89	4.6	7.1
5% Si_3N_4	5	10	85	4.5	8.7
10% Si_3N_4	10	10	80	4.5	10.7
20% Si_3N_4	20	10	70	4.6	9.3
30% Si_3N_4	30	10	60	4.6	10.4
Si_3N_4 ceramic	Si_3N_4: Al_2O_3: Y_2O_3 = 90 : 5 : 5			3.2	18.6

6.3.3 Sliding Experiments

A Si_3N_4 ball was slid against a disk of the Si_3N_4-TiN-Ti composite or Si_3N_4 ceramic using a unidirectional ball-on-disk machine (Implants Sciences Corporation Model 200). The sliding speed was 40 mm/s, the applied load was 0.49 N, and the sliding time was 2 hours. The lubrication fluids used were purified n-hexane, distilled water and anhydrous ethanol. The viscosity of the fluids was water, 0.9 MPa s; ethanol, 1.0 MPa s; and n-hexane, 0.7 MPa s. The sliding tests were also conducted in a dry condition for comparison.

6.3.4 Results and Discussion

Figure 4.80 shows the steady-state coefficient of friction as a function of the content of Si_3N_4 in the disk. The friction of the sintered material was high in the dry condition, in n-hexane and in water. The friction coefficient in these conditions was in the range of 0.49 to 0.84. On the other hand, ethanol reduced the friction effectively. The friction coefficient in ethanol was in the range of 0.14 to 0.17.

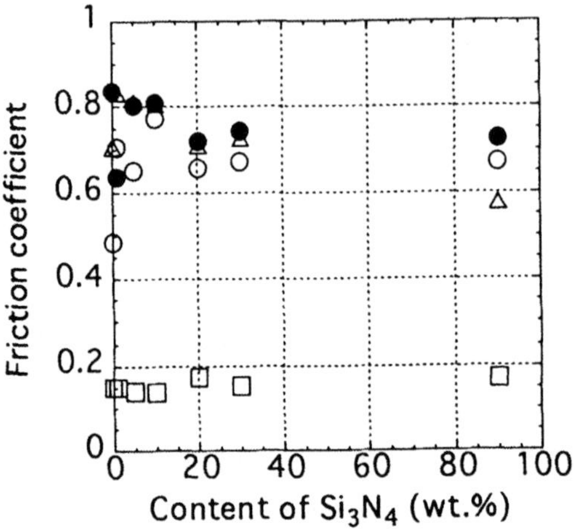

Figure 4.80. Friction coefficient of sintered material against Si_3N_4 as a function of the content of Si_3N_4 in the disk material. (a) Si_3N_4 ball, (b) disk.(● dry, △ n-hexane, ○ water, □ ethanol).

Figure 4.81 shows the specific wear rates (*Ws*) of the Si_3N_4 ball and the disk as a function of the content of Si_3N_4 in the disk. The *Ws* of the Si_3N_4 ball was large in the dry condition, in n-hexane and in water. On the other hand, the *Ws* of the Si_3N_4 ball was much lower in ethanol than in other conditions as shown in Figure 4.81(a). In all tested conditions, the *Ws* of the Si_3N_4 ball had the minimum value when it was slid against the 20 % Si_3N_4 composite.

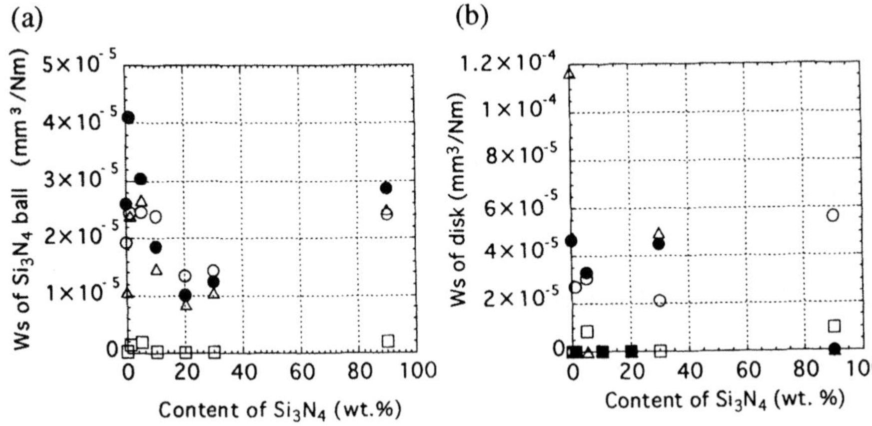

Figure 4.81. Specific wear rate (Ws) of ball and disk as a function of the content of Si_3N_4 in the disk material. (a) Si_3N_4 ball (b) disk(● dry, △ n-hexane, ○ water, □ ethanol).

The *Ws* of the Si_3N_4-TiN-Ti composite disks in water and in ethanol were smaller than that of the Si_3N_4 ceramic as shown in Figure 4.81(b). In the dry condition, the *Ws* of the 0% Si_3N_4 composite, 5 % Si_3N_4 composite and 30 % Si_3N_4 composite were larger than that of the Si_3N_4 ceramic. In n-hexane, the *Ws* of the 0 % Si_3N_4 composite and 30 % Si_3N_4 composite were larger than that of the Si_3N_4 ceramic. In all tested conditions, the *Ws* value of the 10 % Si_3N_4 composite and 20 % Si_3N_4 composite were too small to be measured. As described above, the 20 % Si_3N_4 composite shows good wear resistance in low viscosity fluids and in the dry condition.

Scanning electron microscopy (SEM) images of the worn surfaces of the 20 % Si_3N_4 composite are shown in Figure 4.82. As can be seen in Figure 4.82(a) and Figure 4.82(b), deformation due to plastic flow was observed on the surface of the disk worn in the dry condition and in n-hexane. Elemental analysis using an electron probe micro analyzer (EPMA) revealed that the worn surface mainly consisted of SiO_2. These results indicate that Ti was selectively removed from the sliding surface and that the remaining Si_3N_4 oxidized to form SiO_2 during sliding in the dry condition and in n-hexane. On the other hand, the surfaces of the disk worn in water and in ethanol were smooth as shown in Figures 4.82(c) and 4.82(d). The pits on the surface observed in Figures 4.82(c) and 4.82(d) were introduced during the polishing process. The chemical composition of most of the disk surface did not change during sliding in water and in ethanol. The worn surfaces were partially oxidized in these fluids.

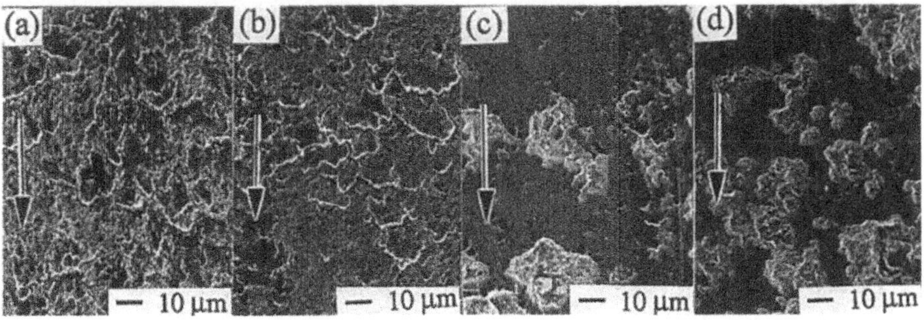

Figure 4.82. SEM images of worn surfaces of 20 % Si_3N_4 composite disk. (a) SEM image of surface worn in the dry condition, (b) SEM image of surface worn in n-hexane, (c) SEM image of surface worn in water and (d) SEM image of surface worn in ethanol. Arrows indicate the sliding direction.

SEM images of the worn surfaces of the Si_3N_4 ball after sliding against the 20 % Si_3N_4 composite are shown in Figure 4.83. On the surfaces worn in the dry condition and in n-hexane, an adhesive substance was observed as shown in Figures 4.83(a) and (b). Furthermore, the presence of Ti in high concentrations on the surfaces of the Si_3N_4 ball worn in the dry condition and in n-hexane was revealed by elemental analysis using energy dispersive spectroscopy. This suggests that adhesive wear occurred on the sliding surfaces between the Si_3N_4-TiN-Ti composite disk and the Si_3N_4 ceramic ball in the dry condition and in n-hexane. Because of the adhesion, friction and wear might be high in these conditions. The worn surface in water was relatively smooth, and flaky wear debris was observed on the edge of the worn surface shown in Figure 4.83(c). As shown in Figure 4.83(d), a powdery substance was observed on the worn surface in ethanol. Ti was present in trace amounts on the worn ball surface in water and in ethanol. Water and ethanol prevented the adhesion of the Ti component to the Si_3N_4 ceramic. In the wear debris on the edge of the surface worn in water, Ti was detected in high concentrations. In water, the Si_3N_4-TiN-Ti composite disk might be grooved by the Si_3N_4 ceramic ball. Therefore, friction and wear in water might be high. In ethanol, a lubrication film consisting of silicon alkoxide and polyoxysilane might be formed by the tribochemical reaction between Si_3N_4 and ethanol [102]. This lubrication film might reduce the friction and wear of the Si_3N_4-TiN-Ti composite/Si_3N_4 ceramic system in ethanol.

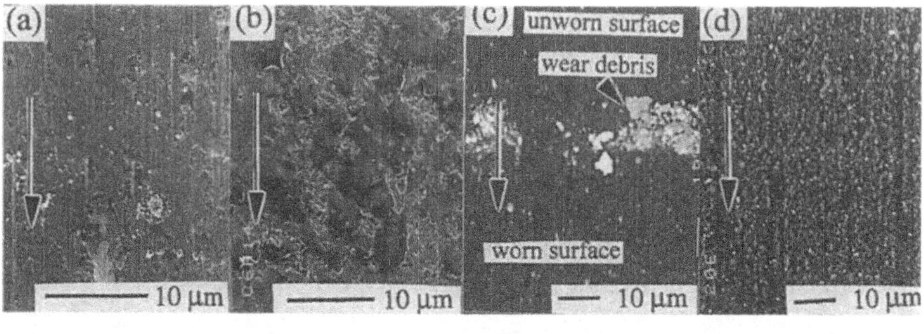

Figure 4.83. Scanning electron microscopy images of the worn surfaces of a Si_3N_4 ball slid against the 20 % Si_3N_4 composite. (a) Si_3N_4 ball worn in the dry condition, (b) Si_3N_4 ball worn in n-hexane, (c) Si_3N_4 ball worn in water and (d) Si_3N_4 ball worn in ethanol. Arrows indicate the sliding direction.

6.3.5 Conclusions

(1) The friction of the Si_3N_4-TiN-Ti composite/Si_3N_4 ceramic system was low in ethanol. On the other hand, in n-hexane and in water, the friction of this system was as high as in the dry condition.

(2) During sliding, the Ti in the Si_3N_4-TiN-Ti composite disk transferred to the Si_3N_4 ceramic ball and resulted in adhesive wear in the dry condition and in n-hexane. Furthermore, the worn surfaces of the composite disk were oxidized in the dry condition and in n-hexane.

(3) Water and ethanol prevent the transfer of Ti from the composites to the Si_3N_4 ceramic. In water and in ethanol, the wear of Si_3N_4-TiN-Ti composite was smaller than that of the Si_3N_4 ceramic
(4) The 20 % Si_3N_4 composite exhibited the best wear resistance of the tested materials.

7. BIOLOGICAL PROPERTIES

7.1 Biocompatibility of Functionally Graded Implants

7.1.1 Introduction

Currently used implants have mostly uniform composition and structure in their longitudinal directions, although the surface is sometimes modified with a coating layer. However implant functions differ from part to part. In the case of a dental implant, the lower part is positioned inside the jaw bone, and the upper is outside the jaw bone and exposed to intra-oral conditions. Therefore the lower part needs more biocompatibility and affinity for bone formation and the upper part needs the mechanical properties and good fit of shape to adapt to prosthetic appliances.

By applying the concept of the FGM structure Watari *et al.* have fabricated functionally graded implants composed of the titanium /hydroxyapatite (Ti/HAP) system through CIP and furnace sintering process[103-106]. However the optimum sintering conditions are quite different for each component. Ti is stable in vacuum or in a reducing atmosphere, and HAP is stable in an oxidizing atmosphere. Ti-HAP is a difficult combination to sinter under the same conditions. The gradient composition had been limited to a range from pure Ti up to 30%HAP at maximum by the furnace sintering method. In this study the spark plasma sintering (SPS) method was applied to fabricate a Ti/HAP FGM implant and its biocompatibility was investigated through animal implantation tests.

7.1.2 Preparation and Observation

Specimens with uniform composition were first prepared at 1100℃, under 40 MPa by SPS (IZUMITECH) after mixed powders of Ti hydrate (SUMITOMO SITIX) and HAP (SUMITOMO OSAKA CEMENT) were packed in graphite mold 20ϕ×(7-14) mm. The FGM specimens were similarly prepared and the gradient composition ranged from pure Ti to

100%HAP, but at a rather lower sintering temperature of 950℃. After sintering was completed, square rods of 1×1×10 mm were cut out for the implants.

The mechanically polished cross sections of the implants were observed before implantation by optical microscopy (OM: OLYMPUS VANOX-S, ZEISS AXIOSKOP 50), SEM (HITACHI S2380N), and elemental mapping was done by XSAM (X-ray scanning analytical microscope, HORIBA XGT2000V) [107]. The properties of density, Young's modulus, Vickers hardness, fracture toughness were measured using uniform specimens of different compositions.

The uniform and FGM specimens were implanted in the bone marrow of rat femora for 4 to 8 weeks and 2 weeks, respectively. Histological observation was done for thinly sectioned and stained specimens in reflection and transmission modes by OM, after the transmission image by soft X-ray was examined.

7.1.3 Experimental Results

The measurement of the properties of the uniform specimens made by SPS showed that the density was very close to the true density and the fracture toughness increased remarkably with the addition of Ti from 0.7(pure HAP) to 3.8(Ti-50vol%HAP)MPa·$m^{1/2}$.

For the FGM specimens made by previous conventional methods, there had been instability of the delayed expansion followed by spontaneous fracture due to the decomposition of HAP and the residual stress. The specimens fabricated by SPS were stable and could be applied for the animal implantation experiments.

Figure 4.84 is an X-ray transmission image of the Ti/HAP FGM implant inserted in the femora of a rat for 2 weeks. The left side of the implant is Ti and the right is HAP. The graded composition can be recognized by the

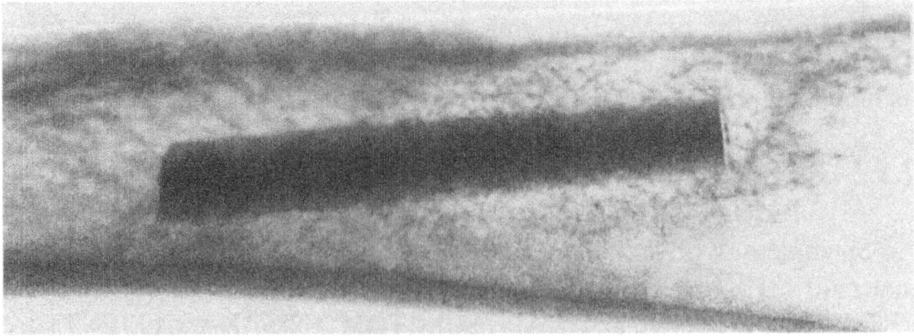

Figure 4.84. X-ray transmission image of Ti/HAP FGM implant inserted in rat femora for 2 weeks. The darker left side of implant is Ti and the slightly brighter right side is HAP.

gradient contrast of the shadow image from the dark portion of a heavier atom Ti to the brighter portion of HAP composed of lighter atoms, Ca, P, O and H. Figure 4.85 shows an image by OM in reflection mode. The left side of the implant is Ti and the right is HAP. The right side with the composition 90 to 100%HAP looks white. Figure 4.86 is a histological image taken in transmission mode and an enlargement is shown for the Ti side(a) and HAP side(b) in Figure 4.87. The new bone was formed surrounding the implant. The newly formed bone layer was observed in direct contact along the surface of the HAP-rich region in Figure 4.87(b).

Figure 4.85. OM image in reflection mode of Ti/HAP FGM implant inserted in rat femora.

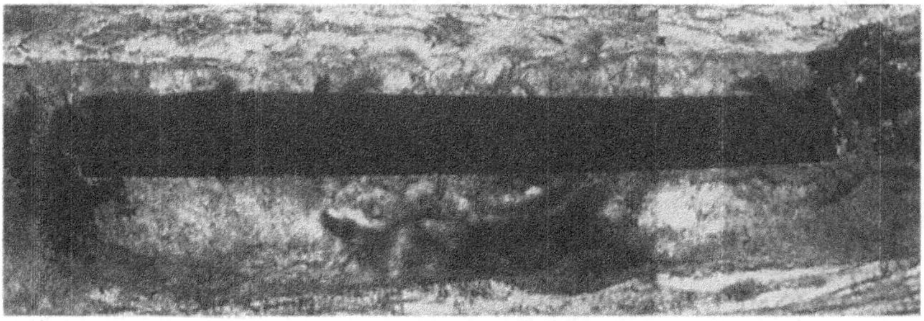

Figure 4.86. Histological image of Ti/HAP FGM implant in rat femora.

7.1.4 Discussion

The most common method for sintering is electric furnace heating after compacting by cold press or cold isostatic press. This needs more than 1300°C for sintering of Ti in vacuum. Under this condition hydroxyapatite becomes unstable and decomposition progresses.

Another method, high-frequency induction heating, is favorable for sintering of Ti because the metal is heated up by itself from the inside and heat is not generated in the ceramics HAP. This may help to avoid the decomposition of HAP. However such specimens are limited to compositions up to 30% HAP since heating is not sufficient for composites with larger amounts of ceramic than 30%HAP.

SPS can heat both the metal and ceramic, and some researchers may also expect further effects such as spark plasma between particles. This may break the surface oxide layer of the Ti, enhance sintering, lower the sintering temperature and reduce the decomposition of HAP.

There was much improvement when sintering by SPS compared with the conventional CIP + furnace sintering method. SEM observation showed that SPS can make more perfectly sintered specimens in spite of the lower sintering temperature. The addition of Ti to HAP had much effect on increasing the fracture toughness and other mechanical properties.

For FGM specimens prepared by conventional methods, the delayed expansion and subsequent spontaneous fracture occurred because of decomposition of HAP and residual stress. In the present study stable specimens could be obtained by lowering the sintering temperature considerably by using the SPS method.

The animal implantation test verified that there was no inflammation and that new bone formed around the implant after 2 weeks. The thin layer of the new bone is formed more in direct contact with the implant surface in the HAP-rich region. Osteogenesis looks pronounced in the HAP-rich region. A longer implantation test is now in progress to derive more reliable results.

A stable Ti/HAP FGM implant was successfully fabricated by SPS and the implant exhibited adequate biocompatibility and affinity for osteogenesis.

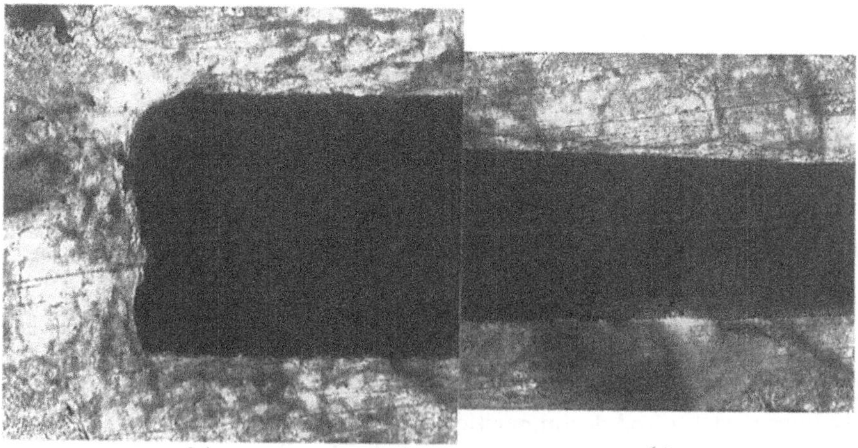

(a) (b)

Figure 4.87. Enlargement of Ti side (a) and HAP side (b) shown in *Figure 4.86*. The newly formed bone layer is more observed in contact with the HAP-rich part.

7.2 Surface Modification of New Ti Alloy for Medical Application

7.2.1 Introduction

The ratio of Japan's population over 65 years old is increasing every year, and the use of artificial body implants is also increasing. Along with various implant materials such as Co-Cr alloy and SUS316L stainless, the use of Ti-6%Al-4%V ELI (Extra low interstitial) alloy is increasing. However, vanadium is classified in the toxic (sterile abscess) group, and aluminum in the capsule (scar tissue) group [108]. A new Ti-15%Zr-4%Nb-4%Ta alloy that contains 0.2%Pd exhibits very much better corrosion resistance, mechanical properties and cytocompatibility than the Ti-6%Al-4%V ELI alloy [109]. In particular, the relative growth ratios of L929 and MC3T3-E1 cells for the new Ti-15%Zr-4%Nb-4%Ta alloy are slightly higher than those of the Ti-6%Al-4%V alloy. Also, the addition of 0.2%O and 0.05%N and heat treatment increased the ultimate tensile strength of the new Ti alloy to 1000 MPa. The total elongation and reduction of area were more than 10% and 50%, respectively [109].

Table 4.10. Chemical composition (mass%) of materials.

Titanium alloys	Zr	Sn	Nb	Ta	Pd	Al	V	Fe	O	N	H	C	Ti
Ti-6%Al-4%V ELI	—	—	—	—	—	6.42	4.19	0.198	0.101	—	0.0052	0.011	Bal.
Ti-15%Zr-4%Nb-4%Ta	15.24	<0.01	3.90	3.92	0.22	—	—	0.022	0.162	0.0480	0.0110	0.002	Bal.

The micro-motion between the bone and the implant gives rise to metallic ion release. This results in the stem loosening thus giving rise to pain. This study examined the corrosion resistance under wear conditions and the surface properties of a new Ti alloy nitrided with high-purity nitrogen gas.

7.2.2 Alloy Specimens

The new Ti-15%Zr-4%Nb-4%Ta alloys containing 0.2%Pd, 0.2%O, 0.05%N and Ti-6%Al-4%V ELI were melted by vacuum arc melting. After β and α-β forging, the alloys were annealed for 7.2 ks at 973 K. Table 4.10 shows the chemical composition of these materials.

7.2.3 Corrosion Resistance under Wear Condition

When the current density measured by anodic polarization testing is low, the corrosion resistance of an alloy is excellent. The anodic polarization tests were conducted in Eagle's MEM (Minimum Essential Medium: mainly composed of NaCl : 6.8 g, $C_6H_{12}O_6$: 1 g, KCl : 0.4 g, $CaCl_2$: 0.2 g, NaH_2PO_4 : 0.115 g, and containing various inorganic salts and kanamycin, etc. in 1 dm^3 of pure water) containing 10% fetal bovine serum, 7.5% NaHCO (2.3 vol%), and 3%L-glutamine (1 vol%) (hereafter called Eagle's medium solution, pH=8.2-8.8) solution. The effect of a frictional load on the anodic polarization properties in Eagle's medium solution was investigated.

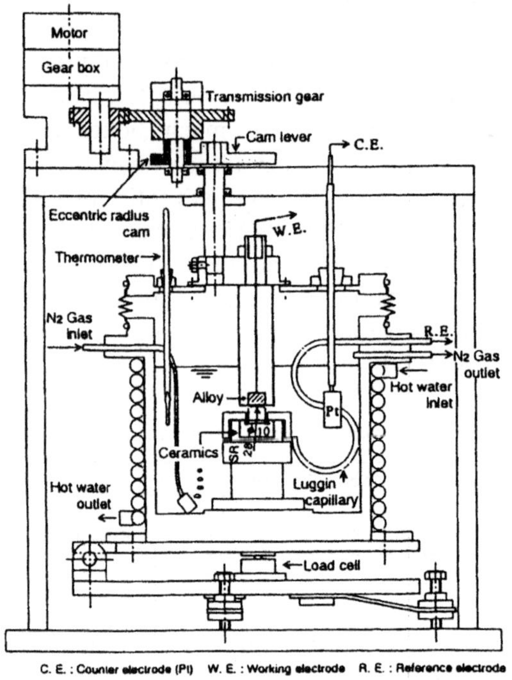

Figure 4.88. Schematic diagram of polarization cell for anodic polarization test under the frictional condition.

The polarization cell used to measure the anodic polarization curves under the frictional condition is shown in Figure 4.88. The specimen electrode is reciprocally moved by a cam while in contact with the apatite ceramic. The applied load was displayed digitally by the load cell and the frictional load was adjusted with screws. To maintain the testing solution at 310 K, hot water was circulated around the cell. The reciprocal speed was varied by changing the gears and the ratio of teeth in the gear box.

The cell is made of Teflon, and the holder to fix the ceramics is made of polychloro tetrafluroethylene (PCTFE, Daiflon).

A test specimen 10 mm in diameter and 10 mm in thickness was prepared from the sample alloy. After connecting the test specimen to a platinum wire by a lug terminal and screw, they were inserted into the Daiflon, and the specimen surface, except for 0.8 cm^2, was covered with epoxy resin. The specimen surface was then polished with waterproof emery paper up to #600 grit under running water, followed by ultrasonic cleaning in ethanol.

The reciprocal distance was 5 mm, apatite ceramic was used for the friction specimen to simulate bone tissue, the data collecting interval was 2 s, the reciprocal speed at 10^{-2} m/s (1 Hz), and the frictional load was up to 59 N. However, because the frictional area in this test was less than 10% of the surface area, the measurement area was assumed to be a constant area of 0.8 cm^2. After the specimen electrode and the friction specimen were set in the polarization cell, the testing solution was deaerated with high-purity nitrogen at a rate of 1.67 x 10^{-6} m^3/s for 0.6 ks. The specimen was initially held at -1.5 V for 300 s. An anodic polarization test was then carried out from -1.5 V to 5 V at a sweep rate of 40 mV/min with a small amount of high-purity nitrogen gas flowing over the solution surface. Moreover, during the anodic polarization test, the measurement was performed by adjusting the screws to minimize frictional load fluctuation (to a maximum 40% of the mean frictional load).

7.2.4 Nitriding Test Using Nitrogen Gas

Test specimens 15 mm in diameter and 10 mm in thickness were cut from the sample alloy. The specimen surfaces were polished with waterproof emery paper up to #1000 grit under running water, followed by ultrasonic cleaning in ethanol and acetone. The Ti alloy specimens were nitrided at 800°C for 6 h, 24 h and 48 h using high-purity nitrogen gas. Micro-Knoop hardness (load: 25 g) was measured toward the center from the alloy surface.

7.2.5 Wear Access Corrosion Resistance for Ti Alloy

Figure 4.89 shows comparisons of anodic polarization curves under a static condition and under the mean frictional load of 49 N at a frequency of 1 Hz with apatite ceramic in the Eagle's medium solution at 37°C. The current density under the frictional condition was higher than that of the static state and fluctuated with destruction and formation of a passive film.

For the SUS316L stainless steel and Co-Cr alloy, the fluctuation of the current density was seen only in the passivity zone, but almost no fluctuation was seen in the active and transpassive zones. As for the Ti alloy, fluctuations of the current density arising from friction were seen from the activity zone to the high-potential region, increasing in the fluctuation widths of the current density (maximum-minimum current densities) and the mean current density (the mean value of maximum and minimum values).

Especially in the case of the Ti-15%Zr- 4%Nb-4%Ta alloy, the fluctuation of the current density was small up to the high-potential zone as shown in Figure 4.89(d).

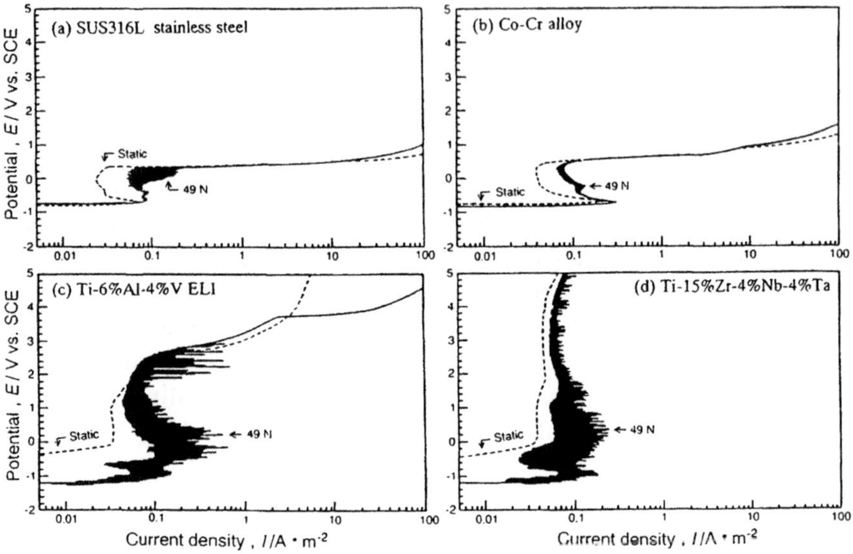

Figure 4.89. Comparison of anodic polarization curves for various implantable alloys in deaerated Eagle's medium solution under the frictional condition at 310 K.

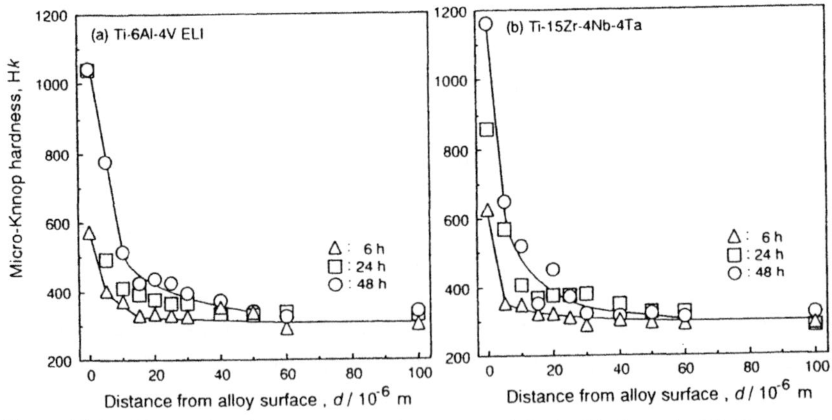

Figure 4.90. Change in micro-Knoop hardness toward the inside from the Ti alloy surface nitrided with high-purity nitrogen gas at 800°C for 6 h, 24 h and 48 h.

7.2.6 Surface Improvement of Ti Alloy by Nitriding

Figure 4.90 shows changes in the micro-Knoop hardness from the Ti alloy surface toward the inside. The micro-Knoop hardness in the alloy surface was high, and it increased with longer nitriding times. The thickness of the nitrided layer was about 15μm in the nitriding test for 48 h at 800 °C.

7.2.7 Conclusions

The current density becomes higher during friction than during static conditions. The fluctuation width was observed in the passivity zone for the Co-Cr alloy. For the new Ti alloy, during friction against the apatite ceramic, the current density was low in the high-potential region, and the fluctuation width of the current density was narrower than for the Ti-6%Al-4%V ELI alloy. The thickness of the nitrided layer was about 15 μm in the nitriding test for 48 h at 800°C.

7.3 Hierarchical and Functionally Graded Structures

7.3.1 Introduction

When developing a new functional material one can learn much from the various smart mechanisms of natural materials, for example, plants and animals, because they acquired many intelligent properties over a long time. One of these is adapted growth which changes their body by themselves to adapt to their environmental conditions through a struggle of existence. The adapted growth triggered the development of a functionally graded structure. Plants and animals also have hierarchical structures that provide different functions and structures on different scale levels. The functionally graded structure added to the hierarchical structure generates far superior functions for these natural materials. Bamboo is a typical example. Understanding natural materials with functionally graded structures is very helpful for developing a new functionally graded material. On the other hand, functionally graded materials have been developed not only in the field of thermal stress relaxation, but also in others [110,111]. The intelligent structure of bamboo can suggest to us a key concept for constructing new FGMs. This paper presents the hierarchical and functionally graded structures of bamboo and shows how bamboo exhibits smart characteristics derived from these structures.

7.3.2 Hierarchical Structure of Bamboo

Bamboo is a hollow cylinder with discretely distributed nodes as shown in the longitudinal cross section of Figure 4.91(b). It is a composite material which is longitudinally reinforced by strong fibers called the vascular bundle sheath as shown in the transverse cross section, Figure 4.91(a). The matrix is filled by soft tissue cells called parenchymatous cells.

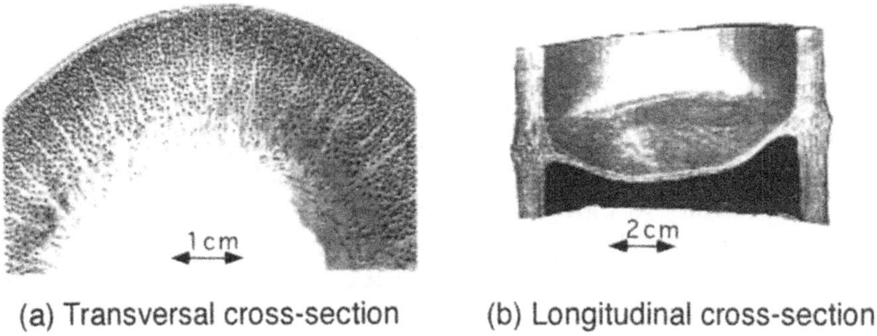

(a) Transversal cross-section (b) longitudinal cross-section

Figure 4.91. Cross section of bamboo culm.

Looking at bamboo based on different scale levels, the features and structure appearing on these levels can be approximately represented by Figure 4.92, which is called the hierarchy [112] of bamboo. The 1st scale level on the bamboo shape, the 2nd scale level on the fiber distribution, and 3rd scale on the fiber-wall structure have different functionally graded structures for the different objects. These multi-functionally graded and hierarchical structures can generate the superior properties of bamboo.

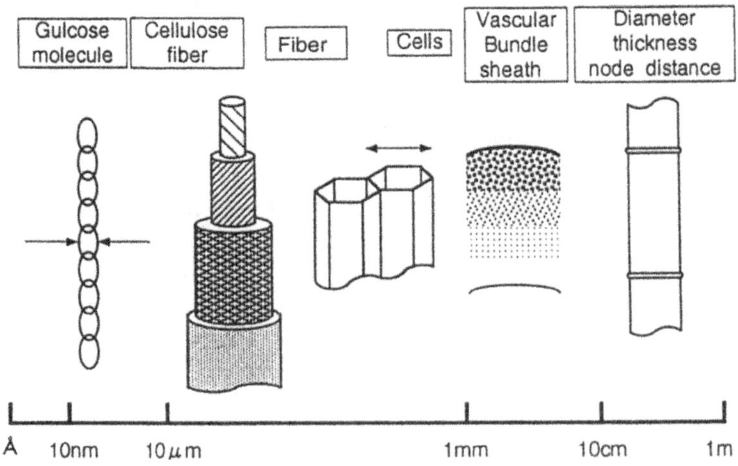

Figure 4.92. Hierarchy of change in bamboo.

7.3.3 Shapes: The 1st (Macroscopic) Functionally Graded Structure

Figure 4.93 shows the distribution of diameter D, thickness t and node length L with respect to the culm number n which is defined by the culm between two adjacent nodes. These parameters specify the feature or shape

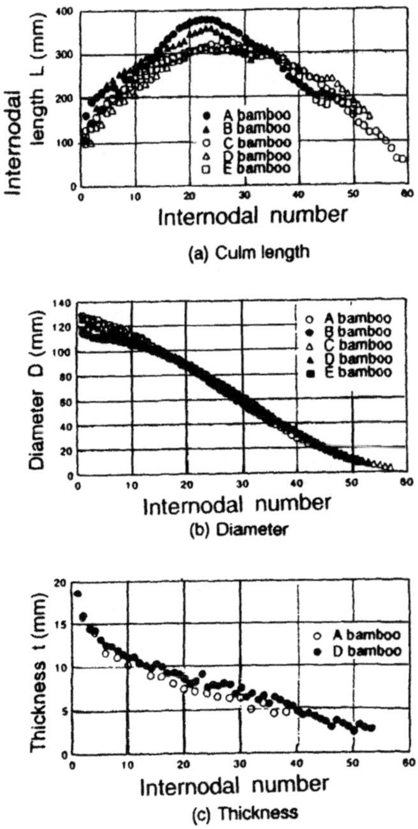

Figure 4.93. Bamboo geometry.

of a particular type of bamboo. D and t decrease monotonically with n, but t varies parabolically, which has its maximum value at the middle point of the total height. These distributions construct a functionally graded structure in the height of the bamboo. Bamboo is mainly subjected to wind loads which induce a bending stress in the bamboo culm. The normalized bending stress on the culm surface is plotted with respect to culm number n(height) in Figure 4.94. The result indicates that the maximum surface stress at every height is almost a constant except for the uppermost region. Therefore, the bamboo culm can be macroscopically classified as a smart structure based on its optimum design [113]. In another words, the 1st graded structure leads to the design of a constant surface stress.

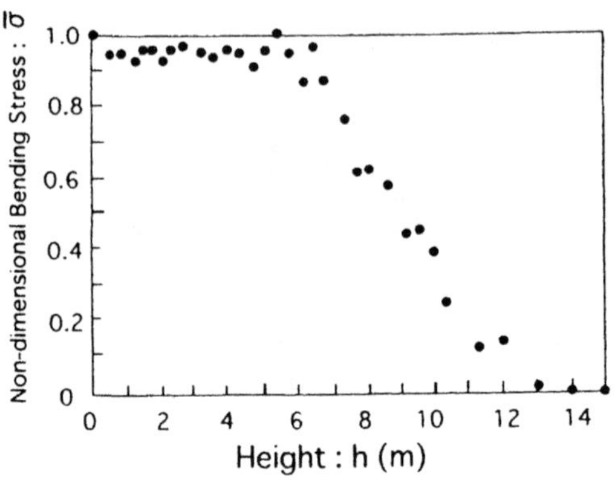

Figure 4.94. Surface stress distribution.

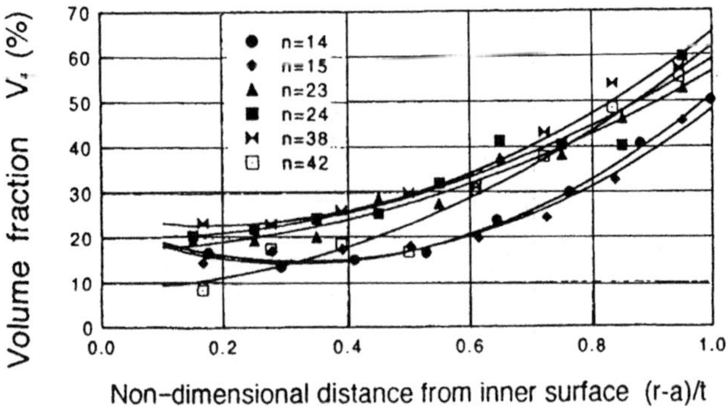

Figure 4.95. Volume fraction of fibers.

7.3.4 Fiber Reinforced Culm: 2nd Functionally Graded Structure

Since the bamboo culm is a fiber-reinforced composite, the fiber distribution may play an important role in the structural property. The volume fraction V_f of fibers was measured in the radial direction and is shown in Figure 4.95. The fiber distributions construct a functionally graded structure, and its maximum value reaches about 60% in the outer surface layer and the lowest value of 10 to 20% in the inner layer.

Young's modulus and the strength of the bamboo culm were measured by tensile tests and the results are shown in Figure 4.96. Applying the mixture principle to these results, can be derived the fiber and matrix properties that are listed Table 4.11. It is remarkable that the fiber strength reaches 610 MPa which is higher than the strength of steel.

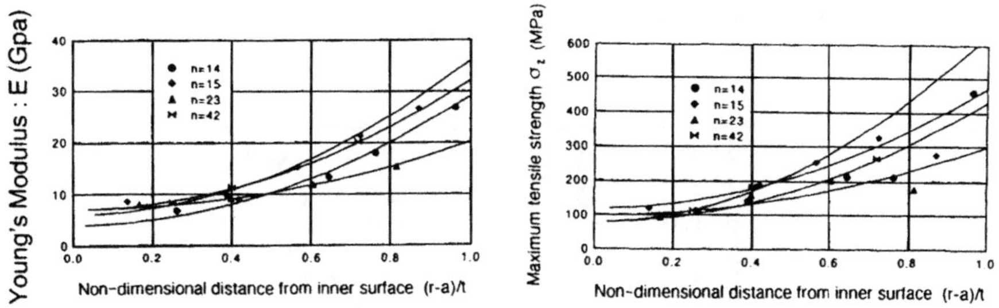

(a) Young's modulus (b) Strength

Figure 4.96. Young's modulus and strength with radius by tensile test.

Table 4.11. Mechanical properties.

Woods	Strength σ_f (MPa)	Young's Modulus E (GPa)	Density ϱ (g/cm^3)
Cedar	29.3–48.5	4.4–9.8	0.29–0.46
Fir	30.7–33.8	5.9–6.7	0.31–0.34
Pine	34.0–41.6	6.5–8.8	0.35–0.42
Spruce	31–40	7.3–8.5	0.38
Hickory	62.5–81.0	8.9–11.4	0.56–0.67
Oak	47.7–74.9	7.9–12.4	0.53–0.61
Bamboo (Fiber)	610	46	1.16
Bamboo (Matrix)	50	2	0.67
Bamboo (Composite)	140–230	11–17	0.6–1.1

The graded structure of the fibers can likely relate to a smart concept. Bamboo is mainly subjected to wind loads which generate bending moment M. Applying the theory of a multi-layered cylindrical beam to the bamboo culm, the bending stress in each layer can be evaluated as shown by the hollow dots of various shapes in Figure 4.97. On the other hand, the normalized measured strength is simultaneously marked by solid dots of various shapes. This distribution is provided by the bamboo to resist the bending stress due to environmental loads. When considering the individual differences of bamboo, both values are very close. In other words, the distribution of fibers adapts to the environmental loads.

7.3.5 Fiber Wall Structure

The bamboo fibers are 10 to 100 μm in diameter depending on the radial locations. The fiber wall consists of multi-layers. The detailed structure was proposed by Liese [114] as shown in Figure 4.98. It consists of four layers and each layer is reinforced by finer fibers called microfibrils and is constructed by filament winding structures. Moreover, the winding angle is

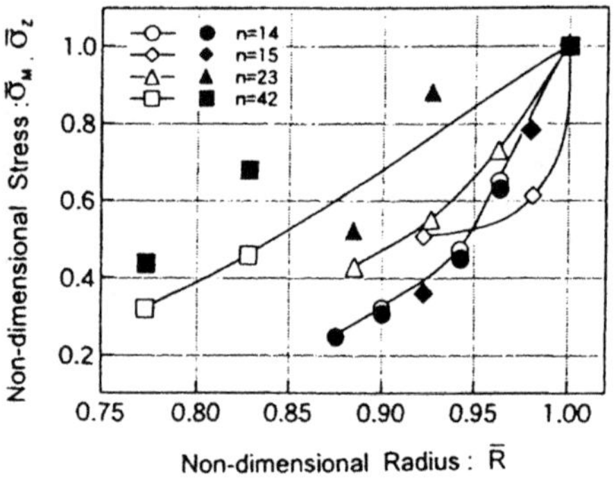

Figure 4.97. Stress and strength distributions.

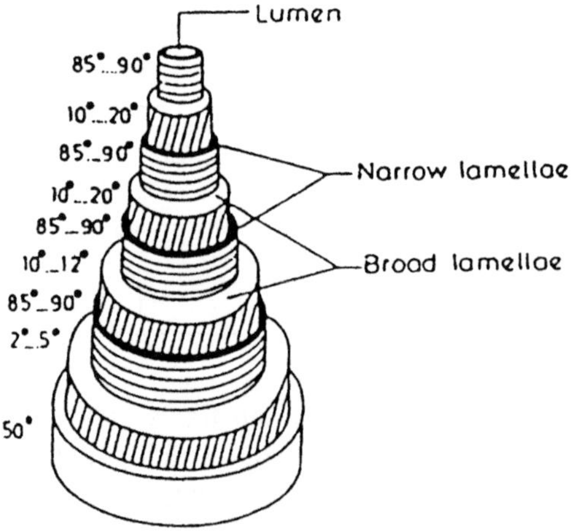

Figure 4.98. Wall structure of bamboo fiber.

different in each layer. This wall structure can be regarded as a functionally graded structure. The question arises for what purpose this structure exists, but there is as yet no clear answer [115].

7.3.6 Conclusions

It was concluded that the superior properties of bamboo are generated by its composite material with strong fibers, and multi-functionally graded and hierarchical structures.

REFERENCES

1. Mihara, T., Asano, S., Obata, M., Watanabe, R. and Kawasaki, A., Elastic Constants of Zirconia/Stainless Steel Sintered Composite Materials Measured by a Line-Focus-Beam Acoustic Microscope, J. Japan Inst. Metals, 56 (1992), 321-326.
2. Makino, A., Araki, N. and Ishiguro, T., Transient Temperature Response in FGM for Step- or Pulse-wise Heating, Ceramic Transaction, Functionally Gradient Materials, edited by Holt, J. B., Koizumi, M., Hirai ,T. and Munir, Z. A., Amer. Ceram. Soc., Westville, Ohio, 305-312.
3. Shinohara, Y., Imai, Y., Ikeno, S., Shiota, I. and Fukushima, T., Thermal Stability of Plasma-Sprayed Ni-Cr-Al-Y/PSZ FGM in Uniform and Gradient Temperature Filed, ibid., 255-262.
4. Aihara, T., Kaji, M. and Igarashi, T., Numerical Analysis of Compositional Distribution Change in Functionally Gradient Materials, J. Japan Inst. Metals, 54 (1990), 758.
5. Takahashi, H. and Hashida, T., Development of an Evaluation Method for Functionally Gradient Materials, Japan Soc. Mech. Eng., Intern. J., Series I, 33 (1990), 281-287.
6. Takahashi, H., Hashida, T., Okukawa, D., Sasaki, M., Hirai, T., Fujii, K. and Kanazawa, M., Oxidation/Thermal Shock Fracture Evaluation Procedure by Plasma-Arc Heating for Functionally Gradient Materials, Proc. 3rd IUMRS Intern. Conference on Advanced Materials, Tokyo, Japan, Aug. 31-Sept. 4, (1993), (in press) .
7. Hashida, T., Takahashi, H., Ishikawa, T., Takahata, K., Kawasaki, A., Watanabe, R., Fukui, K. and Kanazawa, M., Burner Heating Method for Determining Thermal Shock Resistance of Ceramic Coating for Gas Turbine Rotary Blades, Proc. 3rd IUMRS Intern. Conference on Advanced Materials, Tokyo, Japan, Aug. 31-Sept. 4, (1993), (in press).
8. Kawasaki, A. and Watanabe, R., Cyclic Thermal Fracture behavior and Spallation Life of PSZ-NiCrAlY Functionally Graded Thermal Barrier Coating, Proc. of 5th Int. Symp. on Functionally Graded Materials, FGM '98, Dresden, Germany, Oct. 26-29, (1999), 402-409.
9. Takahashi, M., Ito, Y. and Kashiwaya, H., Fabrication and Evaluation of W/Cu Gradient Material by Sintering and Infiltration Technique, Proc. 1st Intern. FGM Symp. 1990, Sendai, Ed. By M. Yamanouchi et al., (1990), 129-134.
10. Koizumi, M., Recent Progress of Functionally Gradient Materials in Japan, Ceramic Engineering & Science Proceedings, The American Ceramic Society, (1992), 333-347.
11. Shimada, M., Yoshii ,T. and Akiyama, S., Ultrasonic Testing of Thermal Shock Damage on Plasma-sprayed FGM Coating, 4th FGM Symposium, FGM Forum Japan, (1991), 31 -35.
12. Nishimori, H., Hayakawa, Y. and Nonaka, T., Nondestructive Evaluation of Functionally Gradient Materials using Ultrasound (2), ibid., 53-60.
13. Saito, M., Takahashi, H., Kawasaki, A. and Watanabe, R.: Acoustic Emission Study on Stress-Induced Transformation and Micro-fractures in Zirconia and Metal Fiber/Zirconia Composites, Proc. The 10th International Acoustic Emission Symposium, Sendai, (1990), 97-104.
14. Takatsubo, J. and Yamamoto, S., Propagation mechanism of ultrasonic waves in porous

ceramics, JSME Int. J., 39-2, (1996), 266-71.
15. Fan, Q., Takatsubo, J. and Yamamoto, S., Quantitative characterization of advanced porous ceramics based on a probabilistic theory of ultrasonic wave propagation, J. Appl. Phys., 86- 7, (1999), 4023-28.
16. Yamanaka, K., Surface acoustic wave measurements using an impulsive covering beam, J. Appl. Phys., 54-8, (1983), 4323-29.
17. Kawasaki, R., *et al.*, Fabrication processes of functionally graded materials, Material, 36-1 (1997), 55-61.
18. Glaeser, A.M., Use of transient FGM interlayers for joining advanced ceramics, Composite Part B, 28 (1997), 1-2, 71-84.
19. Iwasa, S., *et al.*, Study on graded coatings by plasma spraying, Report of Fukui Industrial Technology Center, 6 (1990), 97-99.
20. Kitaguchi ,S., *et al.*, Development of functionally gradient material by low-pressure plasma spraying, Proceedings of plasma spraying conference, 55 (1992), 99-102.
21. Hirai ,T., CVD Processing. MRS Bulletin, 20-1 (1995), 45-47.
22. Yuan ,L., Present status of application of processing for metal/ceramic functionally graded material coatings, Jinshu Rechuli, 1 (1999), 30-33.
23. Yaguchi, H., *et al.*, Development and evaluation of high functional sintered composite materials by HIP diffusion bonding method, Report of Miyagi Industrial Technology Center, 23 (1992), 52-55
24. Iwasa, M., *et al.*, Wear of Hot-pressed aluminum nitride, J. Ceramic Soc. Jap., 94-1 (1986), 188-193.
25. Gregory, C.S. and Miyamoto Y., FGM fabrication by combustion synthesis. MRS Bulletin, 20-1 (1995), 52-53.
26. Omori M. *et al.*, Preparation and properties of ZrO_2(3Y)/Ni FGM, Proceedings of the Third Int. Symp. On Structural and Functional Gradient Materials, (1994), 1-76.
27. Valter, S., Clarke, D.R. and Pompe, W., J., Am. Ceram. Soc.,78-3 (1995), 633- 640.
28. Mizuhara, K., Takahashi, S., Kurokawa, J., Morita, N. and Yoshida, Y., MRS Symp. Proc., 505 (1998), 501-506.
29. Lauer, J. L. and Peterkin, M. E., Can. J. Spectroscopy, 21 (1976), 153-158.
30. Meyer, M. K. and Akinc, M., J. Am. Ceram. Soc., 79 (1996), 938.
31. Meyer, M., Kramer, M. and Akinc. M., Adv. Mater., 8 (1996), 85.
32. Pitman, S. H. and Tsakiropoulos, P., RS Symp. Proc. 364 (1995), 1321.
33. Caplan, D. and Cohen, M., J. Electrochem. Soc., 108 (1961), 438.
34. Suresh, S. and Mortensen, A., *Fundamentals of Functionally Graded Materials*, IOM Communications Ltd., (1998).
35. Erdogan, F., Compp. Eng., 5 (1995), 753-770.
36. Jin, Z. H. and Noda, N., ASME J. Appl. Mech., 62 (1994), 738-740.
37. Gu and Asaro, Cracks in Functionally Graded Materials, Int. J. Solids Structures, 34-1 (1997), 1-17.
38. Erdogan paper which describes increase of KI with change in compliance.
39. Jin and Batra, Some Basic Fracture Mechanics Concepts in Functionally Graded Materials, J. Mech. Phys. Solids, 44-8 (1996), 1221-1235.
40. Suresh, S., Giannakopoulos, A. E. and Olsson, M., J. Mech. Phys. Solids, 42 (1994), 979-1018
41. Winter, A. N., Synthesis of Ni-Al_2O_3 Composites with Compositionally Graded Microstructures, Deformation Analysis with Phase Shifted Moire Interferometry, Ph.D. Thesis, Colorado School of Mines, Golden, CO (1999).
42. Winter, A.N., Steffler, E.D. and Reimanis, I.E., Deformation and Fracture of Ni-Al_2O_3 -Composites Captured with Phase-Shifted moire Interferometry", in press in Ceramic Engineering and Science Proceedings, (2000).
43. Chapa, J. and Reimanis, I. E., Deformation and Fracture of Cu/W Graded Composites, to be published in Acta Materialia, (2000).

44. Steffler, E. D., Frictional Sliding Energy Partitioning of a Center Crack Loaded in Remote Shear and Compression, Ph.D. Thesis, New Mexico State University, Las Cruces, NM, (1999).
45. Chapa, J. and Reimanis, I. E., unpublished work.
46. Oliver, W. C. and Pharr, G. M., An improved technique for determining hardness and elastic modulus using load and displacement sensing indentation experiments, J. Mater. Res., 7-6 (1992), 1564-1583.
47. Certain trade names and products of companies are identified in this paper to adequately specify the materials and equipment used in this research. In no case does such identification imply that the products are necessarily the best available for the purpose or that they are recommended by NIST.
48. Takatsubo, J., National Research Institute, Chugoku, Private communication, (1999).
49. Takenaka, T. and Sakata, K., Jpn. J. Appl. Phys., 19 (1980), 31.
50. Takenaka, T. and Sakata, K., Ferroelectrics, 38 (1981), 769
51. Nishida, I., Symposium Abstract of Chem. Soc. Jpn. (in Japanese), (1997), 145.
52. Liu, S. T. and Long, D., Proc. IEEE, 66 (1978) 14.
53. Kinoshita, Y., Hayashi, N., Watanabe, S. and Ohashi, A. , National Convention Record I.EEE. Japan (in Japanese), 6 (1994), 17.
54. Index of Rocks and their Applications, Committee for the Index of Rocks and their Applications, (in Japanese), Nichia shobo, (1986), 300.
55. U.S. Patent No.5167813, (1992), Japanese Patent No.2947902, (1999).
56. Japan Industrial Standard, JIS -R-8101, (1959).
57. Japan Industrial Standard, JIS -C 2110, (1994).
58. Southwell, W. H., Appl. Opt., 36 (1997), 314.
59. Bovard, B. G. Appl. Opt., 32 (1993), 5427.
60. Perelman, N. A., App, J. Phys., 79 (1996), 2839.
61. Ouellette, M. F., Lang, R. V., Yan, K. L., Bertram, R. W., Owles, R. S. and Vincent, D., Vac. J. Sci. Technol., A9 (1991) , 1188.
62. Donovan, E. P., Vechten, D. V., Kahn, A. D., Carosella, C. A. and Hubler, G. K., Appl. Opt., 28 (1989), 2940.
63. Partlow, D. P. and O'Keeffe, T. W., Appl. Opt., 29 (1990), 1526.
64. Lim, S., Ryu, J. H. and Wager, J. F., Thin Solid Films, 236 (1993), 64.
65. Sainty, G., McFall, W. D., Mckenzie, D. R. and Yin, Y., Appl. Opt., 34 (1995) , 1526.
66. Bulkin, P. V., Swart, P. L. and Lacquet, B.M., J. Non-Cryt. Sol., 187 (1995), 484.
67. Overend, R., Gibson, D. R. and Marshall, R., Vacuum, 43 (1992), 51.
68. Swart, P. L., Bulkin, P. V. and Lacquet, B. M., Opt. Engin., 36 (1997) , 1214.
69. Wang, X., Masumoto, H., Someno ,Y. and Hirai, T., Appl. Phys. Lett., 72 -25 (1998), 3264.
70. Wang, X., Masumoto, H., Someno, Y. and Hirai, T., J. Vac. Sci. Technol., A17 (1999), 206.
71. Pulker, H. K., Appl. Opt., 18 (1979), 1968.
72. Martin, P. J., J. Mater. Sci., 21 (1986), 1.
73. Albertinetti, N. and Minden, H. T., Appl. Opt., 35 (1996), 5620.
74. Martin, P. J., Macleod, H.A., Netterfield, R.P. , Pacey, C.G. and Sainty, W.G., Appl. Opt., 22 (1983), 178.
75. Feldman, A., Farabaugh, E. N., Haller, W. K., Sanders, D. M., and Stempniak, R. A., J. Vac. Sci. Technol., A4 (1986), 2969.
76. Macleod, H. A., *Thin-Film Optical Filters*, 2nd, Edited by Adam Hilger, Techno House, England, (1986), Chap. 8, p. 334.
77. Hirano, T, *et al.*, Proc. 3rd Int. Symp. on FGM, (1994), 633.
78. Teraki, J., *et al.*, Proc. 4th Int. Symp. on FGMs, (1996), 483.
79. Nishio, Y. and Hirano, T., Jpn. J. Appl. Phys., 36 (1997), 5181.
80. Sheng, L. and Xing, D. Y., Phys. Rev., B50 (1994), 1001.

81. Mahan, G. D., *et al.*. J. Appl. Phys., 83 (1998), 4683.
82. Lake, R. and Datta, S., Phys. Rev., B46 (1992), 4757.
83. Yang, C. and Cheh, H. Y., J. Electrochem. Soc., 142(1995), 3040.
84. Zielonka, A. and Fauser, H., Z. Phys. Chem., 208(1999), 195.
85. Limited space prevents us from presenting the references on FGM work. These will be adequately represented in the other papers of this volume.
86. Krushov, M. M., Principles of abrasive wear, Wear, 28 (1975), 69.
87. Archard, J. F., Contact of Rubbing Surfaces, J. Appl. Phys., 14 (1953), 891-898.
88. Greenwood, J. A. and Williamson, J.B.P., The contact of nominally flat surfaces, Proc. Roy. Soc., London, A296 (1966), 300-319.
89. Lee, S. C. and Ren, N., The subsurface stress field created by three-dimensionally rough bodies in contact with traction, STLE Tribol. Trans., 37 (1994), 615-621.
90. Kim, H., Shin, D. S. and Fischer, T. E., Mechanical and chemical aspects in the wear of Alumina, Proc. Jpn. Intl.Tribol. Conf., Nagoya, (1990), 1437-1442.
91. Hsu, S. M., Lim, D. S., Wang, Y. S. and Munro, R G., Ceramic wear maps: concept and method development, Lubr. Engr., 47 (1991), 49-55.
92. Jahanmir, S. and Dong, X., Mechanism of mild to severe wear transition in alumina, J. Tribology, 114 (1992), 403-411.
93. Liang, H. and Fischer, T. E., Effect of grain boundary impurities on the mechanical and tribological properties of zirconia surfaces, J. Amer. Ceram. Soc., 76 (1993), 325-329.
94. Fischer, T. E., Anderson, M.P. and Jahanmir, S., Influence of fracture toughness on the wear resistance of yttria-doped zirconium oxide, J. Amer. Ceram. Soc., 72 (1989), 252-257.
95. Fischer, T. E., Anderson, M. P., Jahanmir, S. and Salher, R., Friction and wear of tough and brittle zirconia in nitrogen, air, water, hexadecane and hexadecane containing stearic acid, Wear, 124 (1988), 133-148.
96. Fischer , T. E. and Tomizawa, H., Interaction of microfracture and tribochemistry in the friction and wear of silicon nitride, Wear, 105 (1985), 22-32.
97. Gee, M. G., The formation of aluminum hydroxide in the sliding wear of alumina, Wear, 153 (1992), 201-218.
98. Takatsu, S. and Umeda, K., Proceedings of NEDO International Symposium on Functionally Graded Materials. Tokyo, Japan, Oct. 21-22, (1999), 173.
99. Tomizawa, H. and Fischer, T.E., ASLE Trans., 30-1 (1986), 41-46.
100. Hibi, Y. and Enomoto, Y., Wear, 133 (1989), 133-145.
101. Alman, D. E. and Hawk, J. A., Wear, 225-229 (1999), 629-639.
102. Hibi, Y. and Enomoto, Y., Wear, 231 (1999), 185-194.
103. Watari, F., *et al.*, Functionally Gradient Dental Implant Composed of Titanium and Hydroxyapatite, in Proc. 3rd Int. Symp. FGM, edited by Ilschner, B., Presse Polytech., Lausanne, (1995) 703–708.
104. Watari, F. *et al.*, Fabrication and Properties of Functionally Graded Dental Implant. Composites Part B 28 (1997), 5-11.
105. Watari, F., *et al.*, Elemental mapping of functionally graded dental implant in biocompatibility test, *Functionally Graded Materials 1996*, edited by Shiota, I. and Miyamoto, Y., Elsevier, Amsterdam, (1997), 749-754.
106. Watari, F., *et al.*, Biocompatibility of Titanium/hydroxyapatite and Titanium/Cobalt Functionally Graded Implants, Functionally Graded Materials 1998, edited by Kayser, W. A., Trans. Tech. Publ., Zurich, (1999), 356–361.
107. Uo, M., *et al.*, Dissolution of nickel and tissue response observed by X-ray analytical microscopy, Biomaterials, 20 (1999), 747–755.
108. Okazaki, Y., Rao, S., Asao, S., Tateishi, T., Katsuda, S. and Furuki, Y., Effect of metallic Ti, Al and V on cell viability. Materials Transactions, JIM, 39 (1999), 1053-1062.
109. Okazaki, Y., Rao, S., Ito, Y. and Tateishi, T., Corrosion resistance, mechanical properties, corrosion fatigue strength and cytocompatibility of new Ti alloys without Al and V, Biomaterials, 19 (1998), 1197-1215.

110. Shiota, I. and Miyamoto, Y., *Functionally Graded Materials 1996*, Elsevier, (1997).
111. Kaysser, W. A., *Functionally Graded Materials 1998*, Trans. Tech. Publ., (1999).
112. Amada, S., Hierarchical Functionally Gradient Structure of Bamboo, Barley and Corn, MRS Bull., 20 (1995), 35-36.
113. Amada, S. and Munekata, T., The Mechanical Structures of Bamboo in Viewpoint of Functionally Gradient and Composite materials, J. Composite Materials, 30 (1996), 800-820.
114. Liese, W., Recent Research of Bamboo, Procd. Int. Bamboo Workshop, Hangzohn, China, IDRAC Canada, (1986), 196-203.
115. Wainwright, S. A., *Axis and Circumference: The Cylindrical Shape of Plants and Animals*, Harvard University Press, (1988).

Chapter 5

APPLICATIONS

1. TOOL MATERIALS

1.1 Introduction

Functionally graded materials (FGMs) are composites consisting of two different materials with a gradient composition. These materials are attracting attention because they can provide new combined functions that surpass the characteristics specific to each element. The objective of this study was to apply the concept of FGMs to tool materials and to develop high-performance tools. To improve the performance of tool materials, FGMs distribute their functionality between the surface with high wear resistance and the inside with high durability. Moreover compressive residual stress is introduced in the surface region. Ikegaya *et al.* developed two types of functionally graded tool materials. One is a functionally graded cemented carbide. This material consists of a titanium-based ceramic surface layer, with a tough cemented carbide core and an intermediate layer with graded composition. High compressive residual stress was introduced in the surface region. This FGM was applied to a cutting insert and was confirmed to have high resistance to wear, breakage and thermal cracking [1]. The other is a functionally graded cemented carbide, sinter-bonded on steel. This FGM is composed of Cobalt with graded cemented carbide layers with a lower Co composition in the surface and a steel substrate. By residual stress control, the residual stress of the cemented carbide layer the near interface was relieved, while a moderate compressive residual stress was introduced in the surface region of the cemented carbide layer. This material also exhibited high wear resistance, durability, and bonding strength [2]. This FGM was applied to an oil well drilling tool called a stabilizer [3]. This paper reports on the fabrication of the FGM cutting insert and the FGM stabilizer, and on the evaluation results.

1.2 Material Design

1.2.1 Functionally Graded Cemented Carbide

Sintered hard materials used for cutting tools include WC-Co cemented carbide, TiCN-Ni Cermet, and thin-film ceramic coated cemented carbide, as shown in Figure 5.1.

	Cemented Carbide	Cermet	Coated carbide
Structure	WC Co 	TiCN Ni 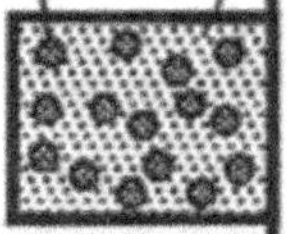	TiCN thin film 
Wear resistance	×	○	○
Toughness	○	×	○
Cost	○	○	×

Figure 5.1. Classification of hard materials used cutting tools.

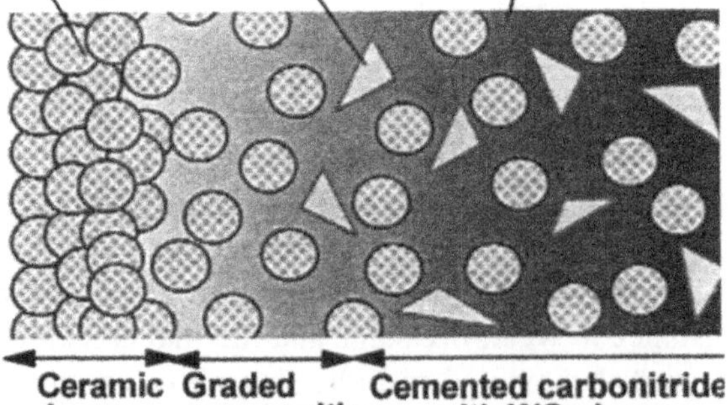

Figure 5.2. Design of functionally graded cemented carbide.

While cemented carbide offers excellent durability, its wear resistance is relatively poor. Conversely, Cermet offers excellent wear resistance, but its durability is poor. Coated materials can provide both excellent durability and wear resistance, but are expensive because of the additional coating process required. To improve cutting performance, functionally graded cemented carbide distributes its functionality between the surface with the high wear resistance of Cermet and the inside with high durability of a cemented carbide, as shown in Figure 5.2. The introduction of compressive residual stress in the surface can be achieved by adjusting the thermal expansion coefficient of the surface so that it is smaller than that of the interior by grading the composition. To suppress manufacturing costs, we aimed to fabricate this FGM by sintering alone using a uniform green compact.

1.2.2 Functionally Graded Cemented Carbide, Sinter-Bonded on Steel

Steel-coated, high wear-resistant hard materials are widely used. A harder and thicker coated layer is required to improve durability. A cemented carbide is a good candidate for this layer. However, it is difficult to coat a hard and thick layer with high bonding strength, because cracks and peeling of the coated layer due to thermal stress are caused by thermal expansion mismatch between the cemented carbide and steel. To overcome this problem and to improve the properties of hard materials, we aimed to control thermal stress and distribute the functionality between the surface with high wear resistance and the inside with high durability. To relieve thermal stress near the material interface and introduce a suitable compressive residual stress in the surface region, high Co composition was selected near the interface and low Co composition was selected for the surface.

1.3 Experimental Procedure

Concerning the functionally graded cemented carbide, TiCN, WC, Co, and Ni powders were combined to a mixture of TiCN, 40% WC, 10% Co and 5% Ni by weight, wet-mixed in a ball mill for 24 hours, and then press-molded to $10 \times 10 \times 5$ mm^3 samples at a pressure of 98 MPa. The samples were heated in a vacuum for sintering to 1673 K and held at that temperature for one hour in a nitrogen atmosphere. The samples were then cooled at slow rate to make the graded structures. To evaluate the structure of the sintered alloy, a cross section of the alloy was observed using optical microscopy and SEM. The hardness distribution was measured using a micro Vickers hardness gauge. EPMA was performed to determine the distribution of the metal binder and X-ray analysis was also used to examine the chemical composition of the component. Residual stress was measured by the X-ray $\sin^2\phi$ method. To evaluate the cutting tool performance, CNMG120408 indexable inserts were fabricated using the same material composition, and then cutting tests were performed.

Concerning the functionally graded cemented carbide sinter-bonded on steel, the WC and Co powders were combined and wet-mixed in a ball mill for 7 hours. After the steel substrate was set in the carbon die, shown in Figure 5.3, the cemented carbide powder was poured onto the steel substrate

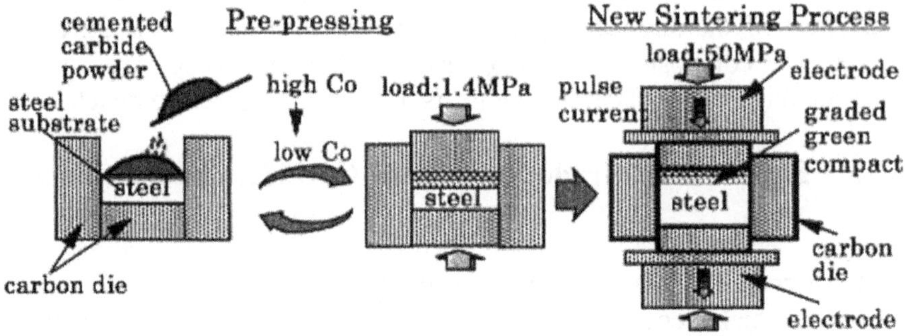

Figure 5.3. Pre-pressing and sintering using new sintering process.

and pre-pressed at 1.4 MPa. This operation was repeated twice first using WC-40% Co and then WC-25% Co powder. The quantity of each powder was adjusted to create a 1.5-mm thickness after sintering. Then the pre-pressed sample was set with the carbon die in the new sintering equipment using pressurizing and direct resistance heating by pulse current, and the sample was sinter-bonded. A 50 MPa load was applied to the pre-pressed sample during sintering. The temperature was set between 1223 to 1273 K on the surface of the carbon die depending on the shape. The evaluation methods were the same as those mentioned above. We applied this FGM to a stabilizer used in oil well drilling to control the forward direction of the drill bit. The outer diameter of the stabilizer is the same as the hole diameter. Because the blade surface rubs and beats against the well wall, the blade surface must be both wear-resistant and breakage-resistant. A conventional welded-blade type stabilizer has three spiral blades with cemented carbide buttons partially inserted, as shown in Figure 5.4. Wear occurs easily on the exposed steel surface where there are no cemented carbide buttons, and the buttons begin to protrude from the worn steel blade, then wear and breakage of the cemented carbide results. To improve the durability, a new stabilizer applying this FGM to the blade, as shown in Figure 5.4 was studied. Three sets of stabilizer blades were welded to the stabilizer steel body. Then the blade surfaces were ground to control the diameter of the stabilizer. To evaluate its performance, a field test was conducted in the MITI Sanriku-oki well in Japan. To compare this FGM stabilizer with a conventional stabilizer, both stabilizers were placed in the same drill shaft. After the field tests, the wear and the breakage of the blades were inspected.

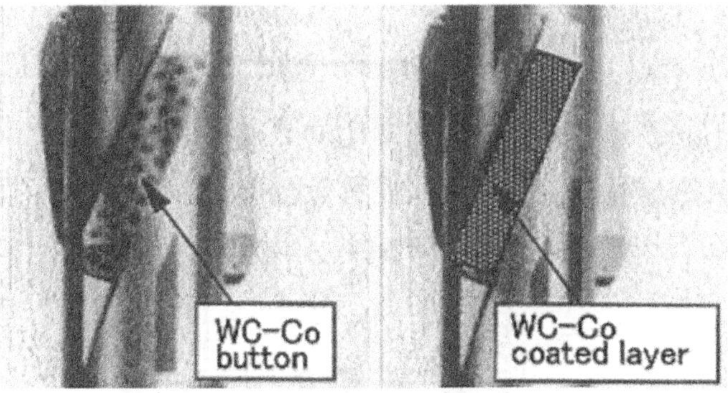

Figure 5.4. Illustration of welded-blade type stabilizer.

1.4 Characteristics of Functionally Graded Cemented Carbide

A SEM micrograph, the hardness distributions and residual stress distributions of the cross section of the specimen are shown in Figures 5.5, 5.6 and 5.7, respectively. X-ray and EPMA analysis confirmed that a functionally graded material with a structure similar to that shown in Figure 5.2 could be obtained by forming a hard Cermet phase mainly comprised of (Ti, W)(C, N) on the surface region, forming a cemented carbonitride phase containing WC in the inside region and then grading the transition between these two phases. The surface hardness was more than 20 GPa in the area

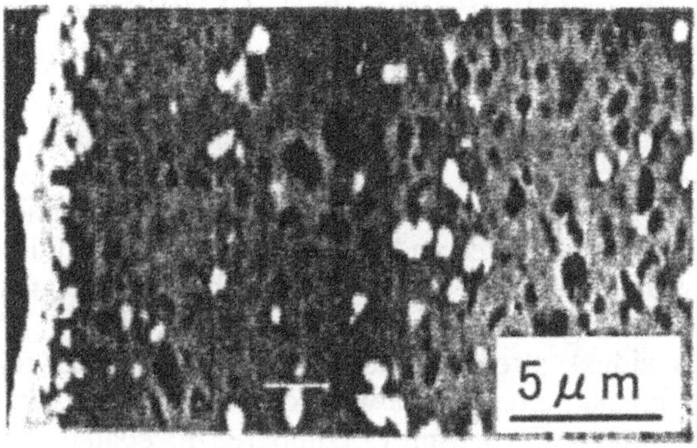

Figure 5.5. SEM micrograph of sample (white areas are WC particles).

within 20 μm depth from the surface and reached a maximum of 22 GPa. High compressive residual stress was introduced in the surface region. The maximum compressive residual stress was as high as 0.8 GPa at 20 μm from the surface.

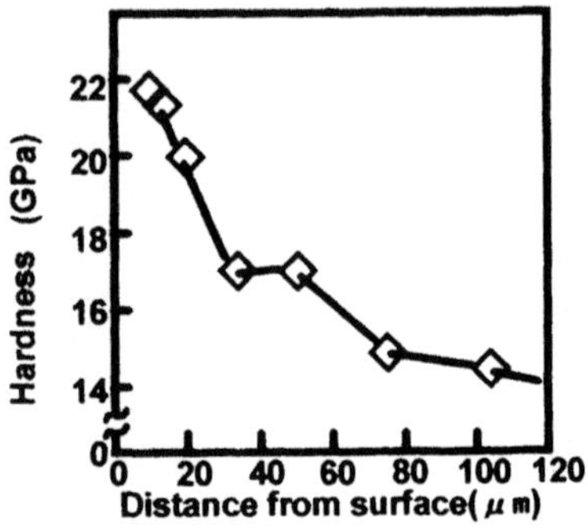

Figure 5.6. Micro-hardness distribution of sample.

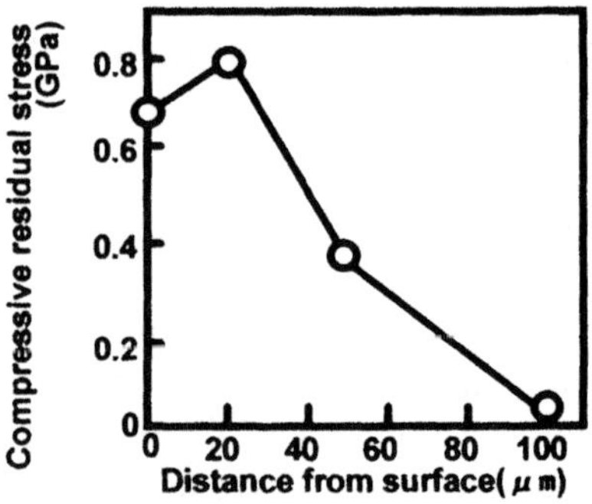

Figure 5.7. Residual stress distribution of sample.

1.5 Cutting Performance of Functionally Graded Cemented Carbide Tools

Using the material design and fabrication techniques described in the foregoing sections, Ikegaya *et al.* developed the functionally graded cemented carbide tool. The performance was evaluated and compared with that of conventional Cermet and coated cemented carbide. The wear and breakage resistance of the FGM tools are shown in Figures 5.8 and 5.9. Because of the high hardness surface layer, tough inner core and compressive residual stress introduced in the surface region, the wear resistance and breakage resistance were confirmed to be superior to those of conventional Cermet. Furthermore, compared to coated cemented carbide, the FGM tools demonstrated cutting performance superior to that of the coated cemented carbide in processes where the peeling of the coated layer occurred easily due to a built-up edge. Peeling does not readily occur in FGM tools due to the continuous transition between the hard surface layer and the internal cemented carbide. It was confirmed that the cutting performance combined the advantages of both conventional Cermet and coated cemented carbides. Moreover, the FGM tools are inexpensive since they do not require the additional coating process.

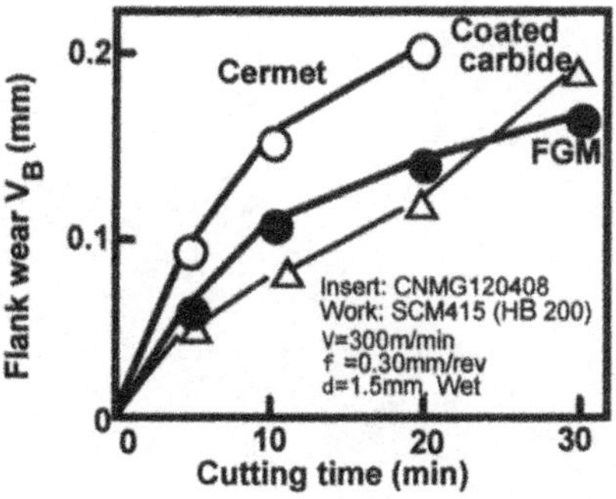

Figure 5.8. Results of cutting wear resistance tests.

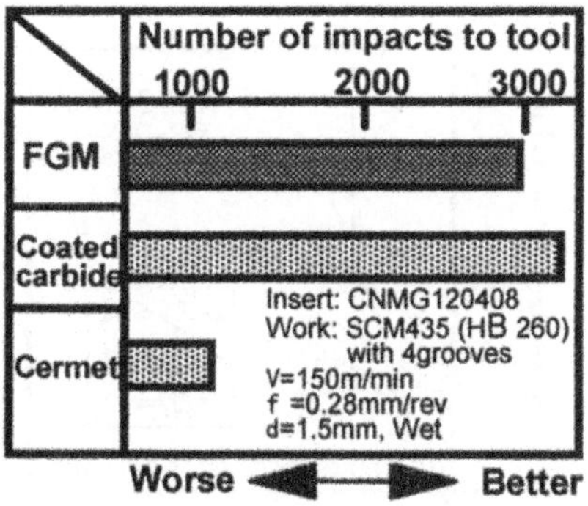

Figure 5.9. Results of cutting durability tests.

1.6 Characteristics of Functionally Graded Cemented Carbide, Sinter-Bonded on Steel

The appearance of the blade after sinter-bonding is shown in Figure 5.10 and suggests that near-net-shape forming was achieved. From observation of the microstructure, the cemented carbide layers were densified completely and few pores were found in them at each point. No cracks were observed near the interfaces between the cemented carbide layer and steel substrate. The hardness distribution of the cross section of the blade was measured and the results are shown in Figure 5.11. The hardness changes sharply in steps corresponding to the graded green compact, and they are nearly the same at various points. The residual stress of the WC-25%Co layer ranges from −0.2 to −0.3 GPa. It is expected that this accounts for suppressed crack initiation and propagation in the cemented carbide layers during drilling.

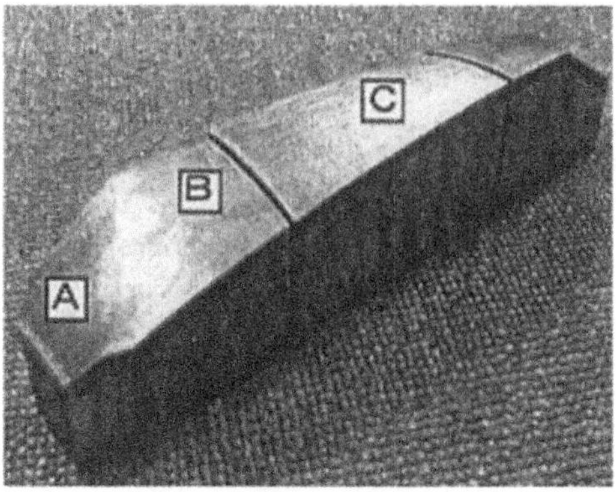

Figure 5.10. Divided stabilizer blades after sinter-bonding.

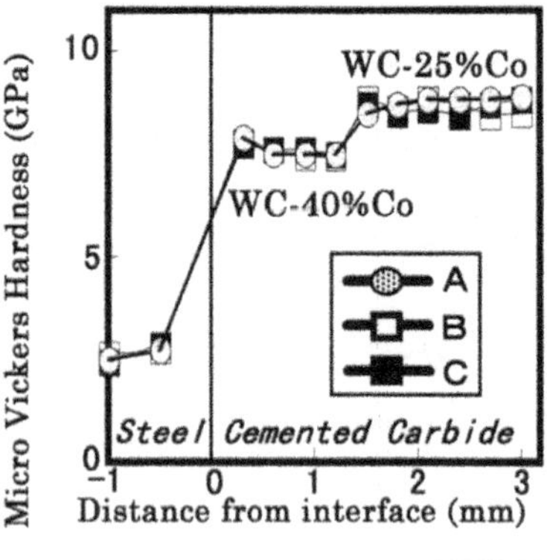

Figure 5.11. Hardness distribution of FGM blades.

After Ikegaya *et al.* made the FGM blades, they fabricated the new stabilizer by a similar process used for fabricating conventional stabilizers. It was confirmed that the conventional fabrication process was also suitable for this new stabilizer [4].

1.7 Evaluation of the FGM Stabilizer in Drilling Tests

To evaluate the performance of the new stabilizer, they carried out field tests at the MITI Sanriku-oki well in an offshore water depth of 857 m. This drilling test was performed between depths of 3520 m and 4500 m (TVD). After 980 m of drilling, neither cracks nor breakage were observed, except a small amount of edge chipping possibly due to mishandling of the FGM stabilizer. In this test, the average wear resistance of the FGM blade was 22 times higher than that of the conventional blade, as shown in Figure 5.12.

This result suggests that the FGM stabilizer can be expected to have high wear resistance when used in drilling deep wells [5]. Preventing chipping due to operator handling may require a more suitable edge treatment.

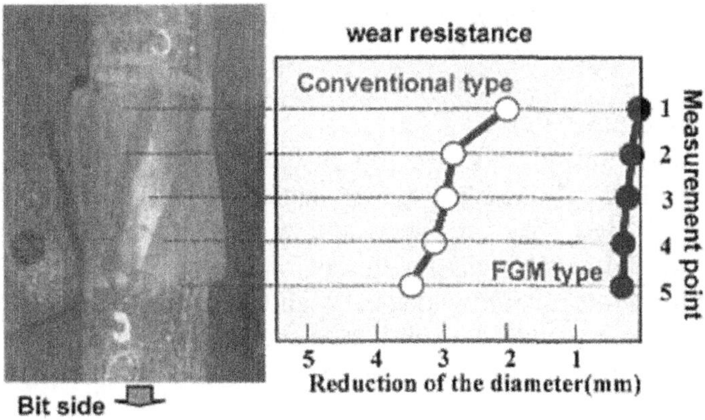

Figure 5.12. Wear of stabilizers at various points.

1.8 Conclusions

Ikegaya *et al.* have applied the material design concept of FGM to the field of tool materials with the objective of improving performance. The FGM distributes its functionality between a high wear-resistant surface and a highly durable interior. Moreover compressive residual stress was introduced in the surface region. They developed two types of functionally graded tool materials. One was functionally graded cemented carbide, and the other was functionally graded cemented carbide, sinter-bonded on steel. Both materials were confirmed to have higher performance than their conventional tool material counterparts. The former FGM has already been put to commercial use in 1996. The potential of high-performance FGM tools was confirmed in this study. Further studies are expected to promote to commercial adaptation and expand the application of these materials.

2. ENGINE COMPONENTS

2.1 Introduction

In recent years, an innovative technique named Spark Plasma Sintering (SPS) to produce new materials called Functionally Graded Materials (FGMs) have been developed in Japan. The nano-instrument indentation technique to evaluate such materials has rapidly progressed in the United States. This method enables the characterization of a localized area of the graded interface of an FGM, which provides important information for FGM design.

The US National Institute of Standards and Technology (NIST) and the Mechanical Engineering Laboratory (MEL) in Japan have been carrying out a 4-year joint research project, from fiscal year 1996 through March 2000, to synthesize and evaluate new functionally graded materials using a spark plasma sintering technique. In Japan, Sumitomo Coal Mining Co., Ltd. (Izumi Technology Co., Ltd.) and Yanmar Diesel Engine Co., Ltd. have joined the project [6, 7].

This joint research focuses on manufacturing a large-size bulk ceramic/metal FGM (FGM/LB)—a combination of a ceramic with high heat-resistance and high wear-resistance and a metal with high mechanical strength. Potential applications for this FGM/LB could provide new wear-resistant materials that could withstand high-temperature environments and which would be suitable for high-efficiency engine materials, ceramic turbine components and mold & die materials for industrial use [8].

The objectives of the joint research are:

1. Development of Heat-resistant FGM/LB Processing Technology
2. Development of FGM/LB Manufacturing Systems
3. Establishment of FGM/LB Evaluation Technology

The project plan also included the development of techniques for the characterization of the FGM properties suitable for understanding the processes and for further development, and to relate the laboratory results of physical and mechanical properties to wear behavior.

2.2 Heat-Resistant FGM/LB Processing Technology

Spark plasma sintering (SPS) is a pressure-assisted pulsed electric current sintering process that has recently been attracting attention. It provides excellent features for use in creating new materials such as composite ceramics and FGMs. The history of SPS technology related to the process in question started in the late 1930s when a sintering process using electrical energizing was introduced in the United States. In Japan, a similar process based on a pulse current applied sintering method was researched and patented in the 60s and is known as spark sintering. This process was not put to wide use due to the lack of application technology at the time, the limited fields where it could be applied and the unsolved problems associated with industrial production, equipment cost and sintering efficiency. There was little literature on research into this process until the latter half of the 70s. The second-generation process was developed from the middle of the 80s to the early 90s. These units were small experimental systems—Plasma Activated Sintering (PAS) with maximum sintering pressures of around 50KN and pulse generators of up to 800 amp—and were used primarily for materials research. From the early 1990s through 2000, the Spark Plasma Sintering (SPS) process became the third generation of this advanced technology—SPS systems with large DC pulse generators of 100 to 3,000 KN incorporating materials-handling robotic mechanisms and 2,000 to 20,000 amp. Moreover, they are considered the most advanced industrial processes for synthetic processing of functionally graded and composite materials. Figure 5.13 shows the basic concept of FGM/LB development for industrial applications. To establish practical use of FGMs using SPS technology, simultaneous research and development on processing software and hardware systems are essential for success in this field.

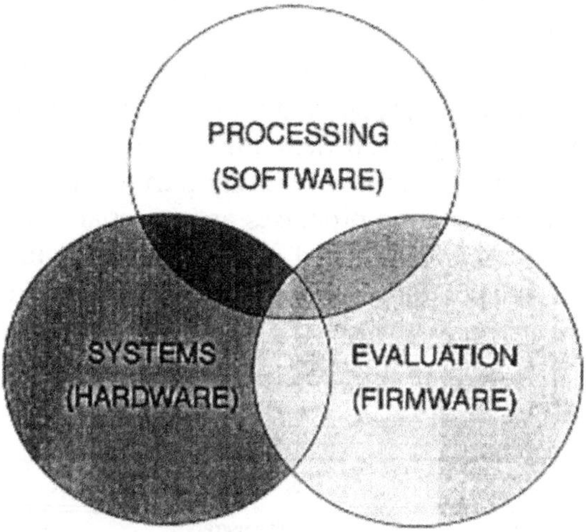

Figure 5.13. Basic concept of FGM/LB development for industrial applications.

2.3 Mechanism of SPS Processing

Conventional electrical hot press sintering processes use DC or commercial AC power, and the main factors causing sintering in these processes are the Joule heat generated by the power supply (I^2R) and the plastic flow of the materials due to the application of pressure. However, the SPS process is an electrical sintering technique which directly applies an ON-OFF DC pulse voltage and current from a special pulse generator to a powder of particles. SPS is regarded as a rapid sintering method, using the self-heating action from the powder's interior, similar to self-propagating high temperature synthesis (SHS) and microwave sintering.

2.3.1 DC Pulse Current Energizing Effect

Figure 5.14 shows a typical ON/OFF DC pulsed current path using conductive powder materials and a sintering die and punch made of graphite. Figure 5.15 illustrates how the pulse current flows through the powder particles inside the SPS sintering die. In addition to the factors causing sintering described above, also another factor is the effective discharge between the particles of the powder occurring at the initial stage of the pulse energizing. When electrical discharges occur, a high-temperature sputtering phenomenon generated by the spark plasma and spark impact pressure eliminate adsorptive gas and impurities from the surface of the powder particles. The action of the electrical field causes high-speed diffusion due to the high-speed migration of ions.

These effects are summarized as follows: The ON-OFF DC pulse-energizing method generates (1) spark plasma, (2) spark impact pressure, (3) Joule heating, and (4) an electrical field diffusion effect. In the SPS process, the powder particle surfaces are more easily purified and activated than in conventional electrical sintering processes and material transfers at both the micro- and macro-levels are facilitated, so a high-quality sintered compact is obtained at a lower temperature and in a shorter time than with conventional processes.

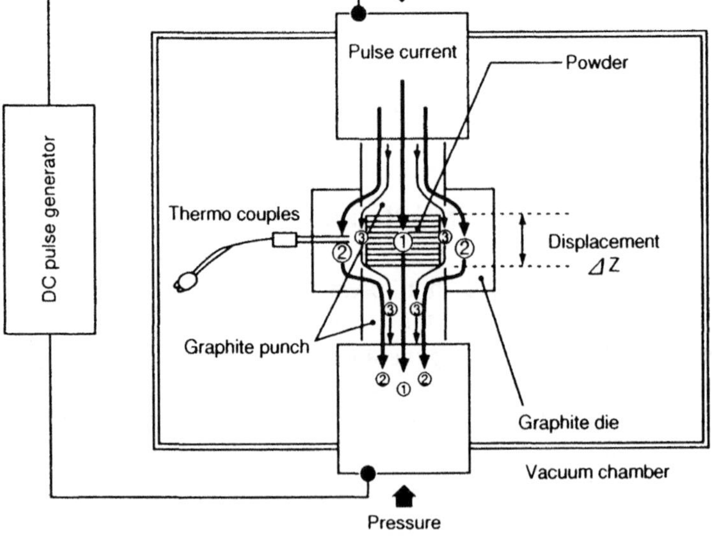

Figure 5.14. ON-OFF pulsed current path.

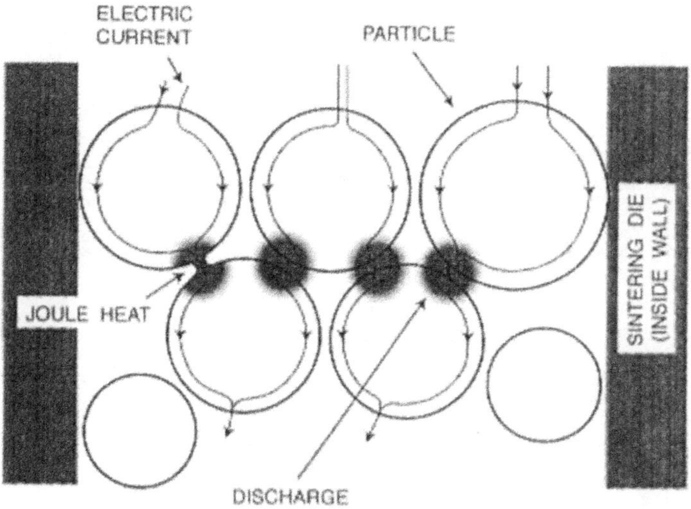

Figure 5.15. Pulsed current flow through powder particles.

2.3.2 Neck Formation by SPS

When a spark discharge occurs in the gaps or at the contact points between the particles of a material, a local high-temperature state (discharge column) of several to tens of thousands of degrees centigrade is generated momentarily. This causes evaporation and melting on the surface of the powder particles in the SPS process, and "necks" are formed around the areas of contact between the particles. Figure 5.16 shows the basic mechanism of neck formation by spark plasma.

Figures 5.17, 5.18 and 5.19 are SEM micrographs showing the results of SPS experiments performed at normal atmospheric pressure with a sintering die and punch made of graphite and a spherical bronze alloy powder. Figure 5.17 shows the behavior in the initial stage of neck formation due to sparks in the plasma. The heat is transferred immediately from the center of the spark discharge column to the spherical surface and is diffused so that the inter-granular bonding portion is quickly cooled. As seen in Figure 5.18, which shows several necks, the pulse-energizing method causes spark discharges one after another between the particles. Even with a single particle, the number of positions where necks are formed between adjacent particles increases as the discharges and/or Joule heating continue.

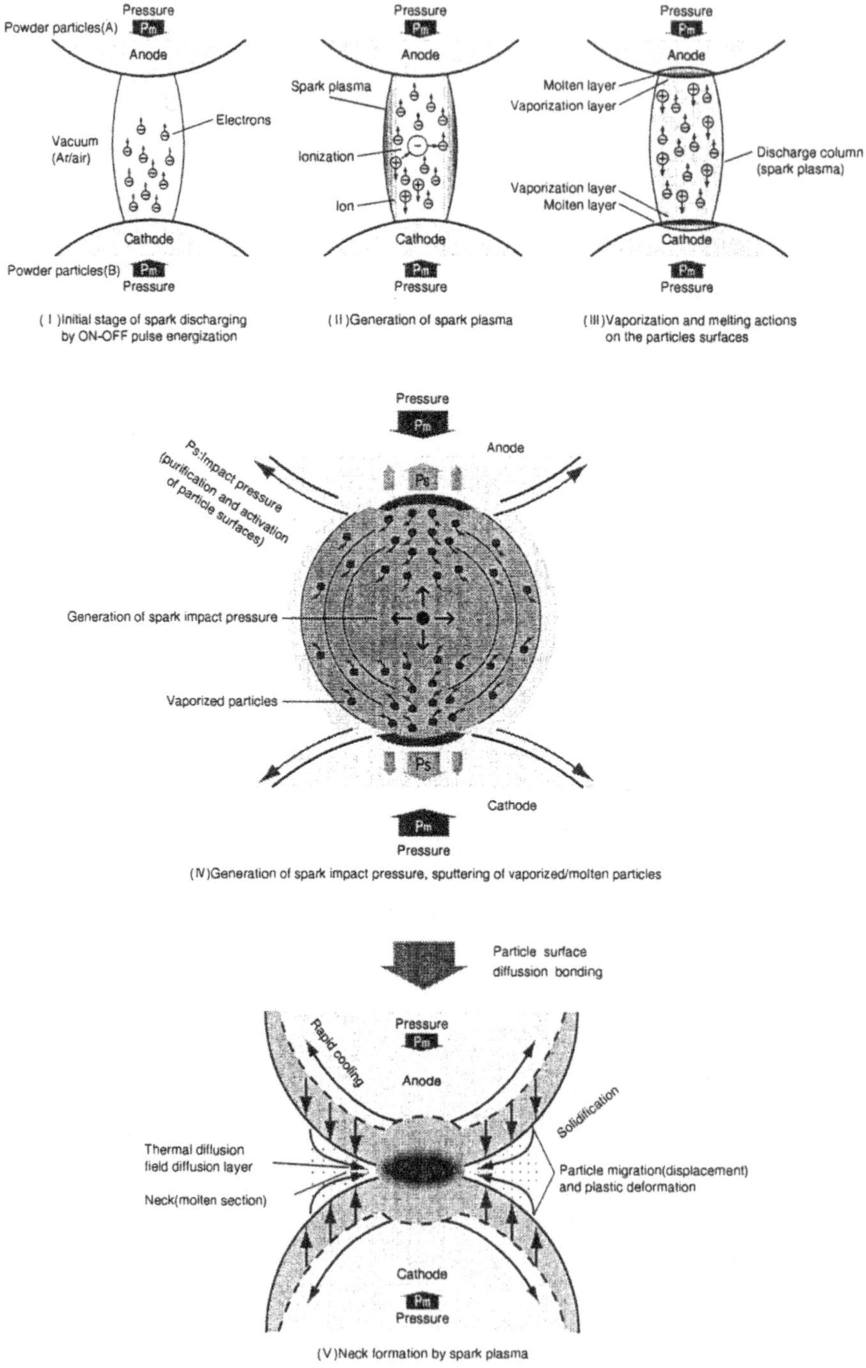

Figure 5.16. Basic mechanism of neck formation by spark plasma.

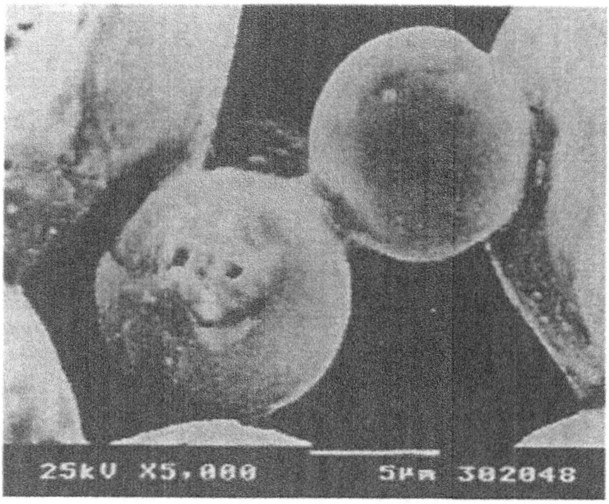

Figure 5.17. Initial stage of neck formation.

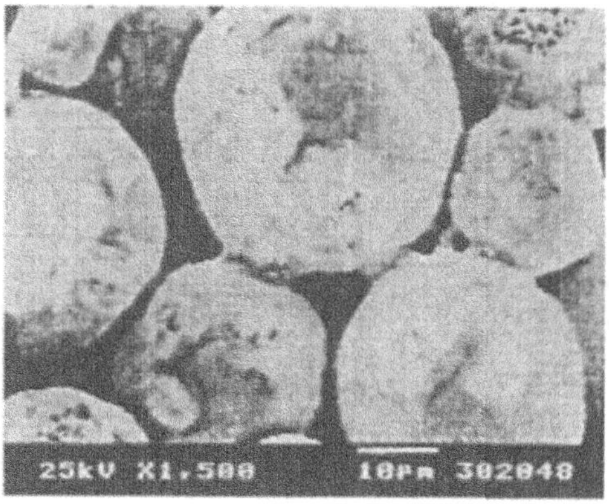

Figure 5.18. Expansion of neck.

Figure 5.19 shows the condition of an SPS sintered grain interface which is plastic-deformed after the sintering has further progressed. This state is the result of processing conditions in which the applied pressure was 29 MPa, the sintering temperature was 773K, the holding time was 120 sec., SPS current 850 amp, and the voltage was 3.9 V.

The graphite sintering dies and punches are subject to Joule heating according to the progress of the sintering of the internal powder material, and they function as heating elements and also maintain the uniformity of the sintering temperature for densification. Figure 5.20 is a comparison of surfaces produced by spark plasma sintering (left) and conventional sintering (right) in a vacuum. Due to the high temperature and sputtering effect of the spark plasma and impact pressure by SPS, rough surfaces were observed.

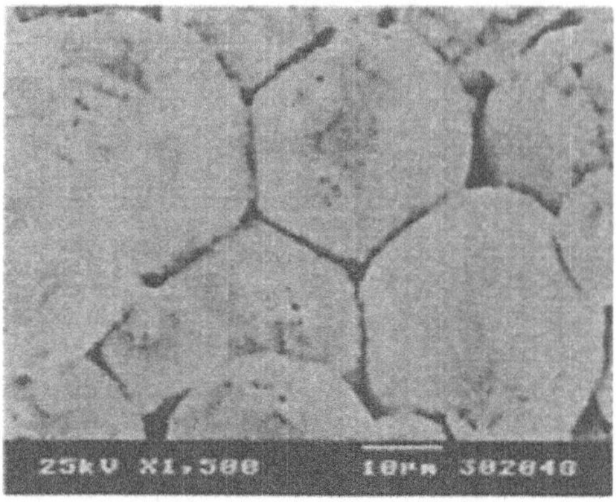

Figure 5.19. Start of plastic flow.

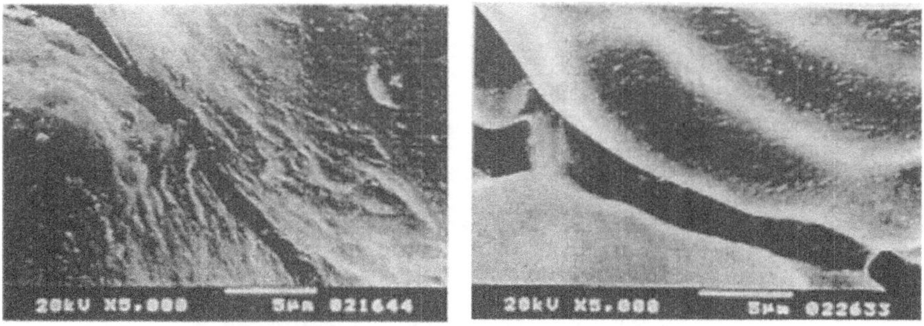

Figure 5.20. Neck formation after spark plasma sintering (left) and normal sintering (right).

2.4 FGM/LB Manufacturing Systems

2.4.1 Automatic FGM Powder Stacking Equipment

An automatic materials-handling robot system and an FGM powder stacking system for the SPS process were developed to manufacture large-size ceramic/metal bulk FGMs. To assure uniform stacked powder thickness over a large area, the FGM powder stacking and lamination are performed by the volume method. As shown in Figure 5.21, this method levels the stacked powder surface in the graphite die cavity with the bottom of the container. The modular configuration uses material powder containers of different volume fractions arranged in a line for FGM/LBs.

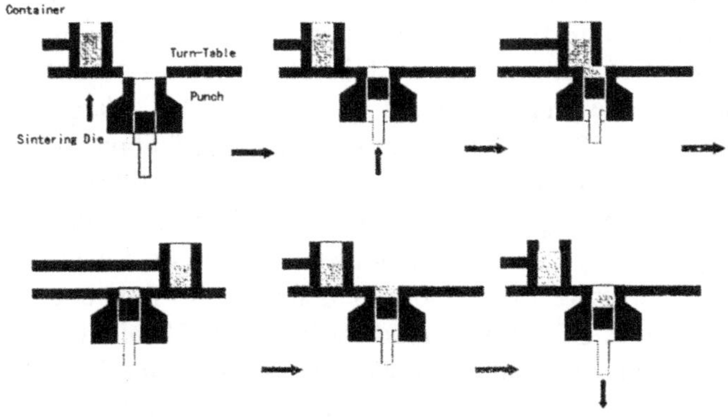

Figure 5.21. Principle of automatic FGM powder stacking.

2.4.2 Fully Automated FGM/LB Manufacturing Systems

Fully automated FGM/LB manufacturing systems were also developed as shown in Figure 5.22. The tunnel type SPS system consists of automatic FGM powder stacking, preheating, SPS sintering, cooling and die release stages. The system employs an automatic materials conveyer, DC pulse power supply, and automatic control systems.

Figure 5.22. Fully automated FGM/LB manufacturing systems.

2.5 FGM/LB Fabrication and Evaluation

Utilizing advanced SPS Systems, $\phi 100$ mm and $\phi 150$ mm ZrO_2

(3Y)/SUS410L stainless steel FGM/LBs were fabricated homogeneously, free of cracks, delaminations or distortion. As an example, the ϕ100 mm, 17-mm thick FGM/LB was sintered in 58 minutes—consisting of a 28 minute heat-up time and 25 minute cooling time by a large SPS system apparatus. Figure 5.23 shows the compositional distribution and Figure 5.24 is the external view of an FGM/LB with a diameter of 100 mm fabricated by SPS. An optical micrograph of the polished surface of the FGM/LB is shown in Figure 5.25. Use of this automated SPS System made possible the automated manufacture of a 150-mm diameter, 15-mm thick ZrO_2(3Y)/ stainless steel FGM/LB with 11 gradient layers in 54 minutes. To evaluate the mechanical, physical and tribological properties of the FGM/LB, various characteristics shown in Figure 5.26 such as Vickers hardness, elastic modulus, tensile strength, friction coefficiency, residual stress, lattice parameter, thermal shock strength, and microstructure were investigated.

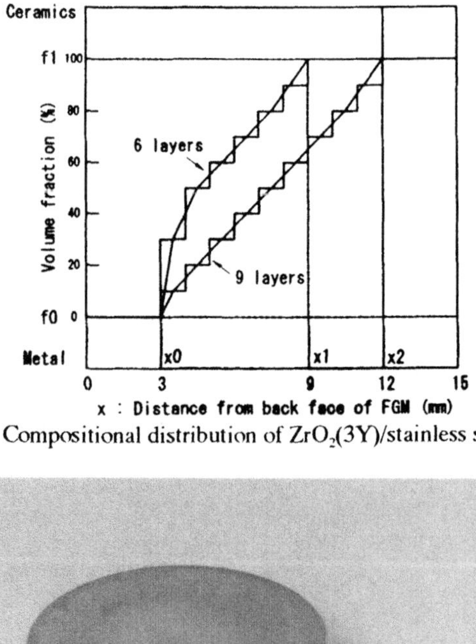

Figure 5.23. Compositional distribution of ZrO_2(3Y)/stainless steel FGM/LB.

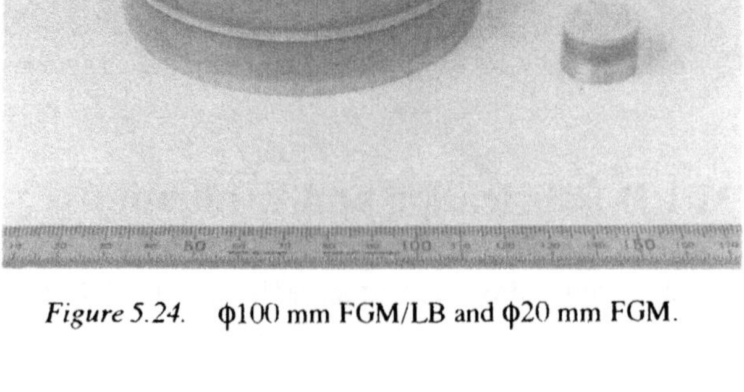

Figure 5.24. ϕ100 mm FGM/LB and ϕ20 mm FGM.

Figure 5.25. Optical micrograph of a fine FGM/LB (11 layers).

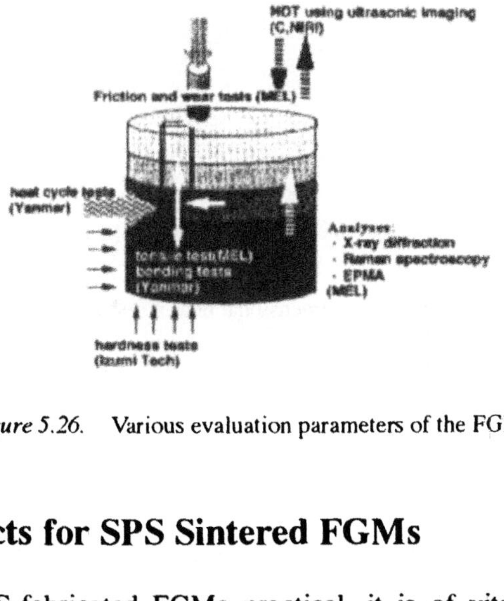

Figure 5.26. Various evaluation parameters of the FGM.

2.6 Prospects for SPS Sintered FGMs

To make SPS-fabricated FGMs practical, it is of vital importance to develop technology to form a 3D Near-Net-Shape as shown in Figure 5.27. Such technology is essential for significantly reducing the processing steps and overall cost of the manufacture of materials for engine components.

Moreover, large-size bulk FGM/LB manufacturing technology has wide-ranging applications in fields other than for automobile materials. For example, the FGM/LB has shown a strong potential for the stamping die and cutting tool markets. As shown in Figure 5.28, WC/Co cobalt compositionally graded cemented carbides make it possible to fabricate new FGM hard alloys that have both high hardness and toughness values. Along with the scaling-up of the production size, such SPS-fabricated compacts are suitable for the manufacture of sputtering target materials, large-size high-quality fine ceramics, and nano-phase bulk materials for related industries. It also makes it possible to expand applications for the electric vehicle and environmental recycling sectors. Therefore, in the near future, the manufacturing technologies for FGMs using the SPS method will open up new markets for advanced materials.

Figure 5.27. Example of 3-dimensional near-net-shape formed by SPS (right).

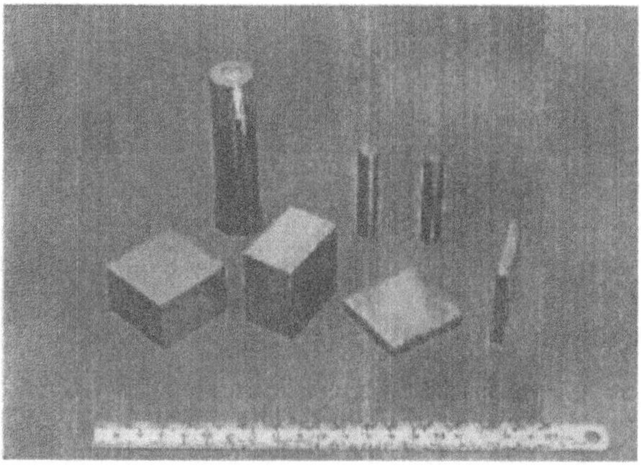

Figure 5.28. Examples of FGM hard alloys fabricated by SPS.

3. SPACE PLANES

3.1 Introduction

A little over a decade ago, the concept of functionally graded materials (FGMs) was conceived as a potential ultra-thermo-resistant material for a thermal protection system (TPS) which was considered the key to developing space planes.

In 1987, the Science and Technology Agency organized a project under a national budget to develop functionally graded materials for the relaxation of thermal stress. Later, since 1993, a new research program Study of the Development of Energy Conversion Materials through the Formation of Gradient Structures promoted research on energy-converting FGMs that remarkably improved energy efficiency [9]. A way to incorporate the positive aspects of the results obtained from these projects into the design of new space planes will be considered.

3.2 Why Must Costs be Reduced One-Hundredth

The international trend has been toward the development of a Re-usable Launch Vehicle (RLV) which has rocket engines as the main source of propulsion. In Japan the controversy continues between different groups favoring two different types of RLV: Single Stage to Orbit (SSTO) and Two Stages to Orbit (TSTO). The SSTO has one body to reach space while the TSTO has two separate parts. The TSTO carries the rocket engine and is considered the easier technical achievement; and for that reason, the author will discuss the potential of cost reduction using TSTO[10]. The goal is to reuse the bodies of this vehicle a number of times; in fact, at least a few hundred times, if possible a few thousand recycles will be attempted. The engine will also be reused hundreds of times, and the purpose is to reduce

Figure 5.29. Body image of TSTO.

the cost to one-hundredth. The development of such a TSTO will be completed in the latter half of the 2010s. Figure 5.29 shows the body image.

However, reduction of launching cost to one hundredth means that higher cost-performance demands for this project are required. This project will be purposeless unless there will be a demand one hundred times what we have now. The most promising among the potential demands for such recycled space planes is the consolidation of the international information network because it necessitates the launching of many satellites. Also promising is the demand for transporting building materials for power plants in space.

A scheme for space-based solar energy plants was formed in response to the global need for an alternative energy source that can replace fossil fuels and nuclear power plants[11]. To construct such power plant in space, twenty to thirty thousand tons of building materials would have to be launched from the Earth into space. Thus it will be necessary to reduce the present cost of launching space vehicles to one hundredth, because the energy must be provided for the same price at the current value. This would require a momentous reformation of our technology.

3.3 Principal Accomplishments of FGM Research Programs

Niino *et al.* have tried to develop an engine for the HOPE-X, an experimental space plane, by applying the FGM concept with a view to higher durability and performance. So far they have developed two engines, the Orbiter and Maneuvering System (OMS) Engine [12] to obtain propulsion in orbit, and the Reaction Control System (RCS) Engine [13] to control the position of the plane.

3.3.1 OMS Engine for HOPE

To cool this engine, they use a combination of a regenerative cooling system (which goes through the passage built axially to the combustion chamber to cool off the chamber wall) and a film cooling system (which is a part of the liquid fuel sprayed on the internal wall of the combustion chamber to cool it off). The fuel used as the filmed coolant does not contribute to propulsion, so to achieve the higher performance of the engine, the amount of filmed coolant must be reduced. To that end, a partially stabilized zirconia (PSZ) coating method has been applied. PSZ is the material used for the thermal barrier, and by putting the layer of PSZ on the internal wall surface of the combustion chamber, the filmed coolant is not necessitated, and at the same time, the temperature of the coolant is maintained below pyrolysis. To relax the thermal stress in this thermal barrier coating, PSZ was gradually applied to a substrate consisting of Nickel and Chromium to form PSZ/NiCr FGM.

Figure 5.30 shows the result of the combustion test, which was repeated 250 times, up to a length of 1,230 seconds in total. The specific impulse that shows the combustion efficiency was 315 seconds, the best in the world in this class. After combustion, delamination of the PSZ layer was observed at the throat where the greatest thermal loading occurs. However, there was no extraordinary increase of the temperature of the regenerative coolant during the combustion test, thus proving that heat resistance and safety can be guaranteed.

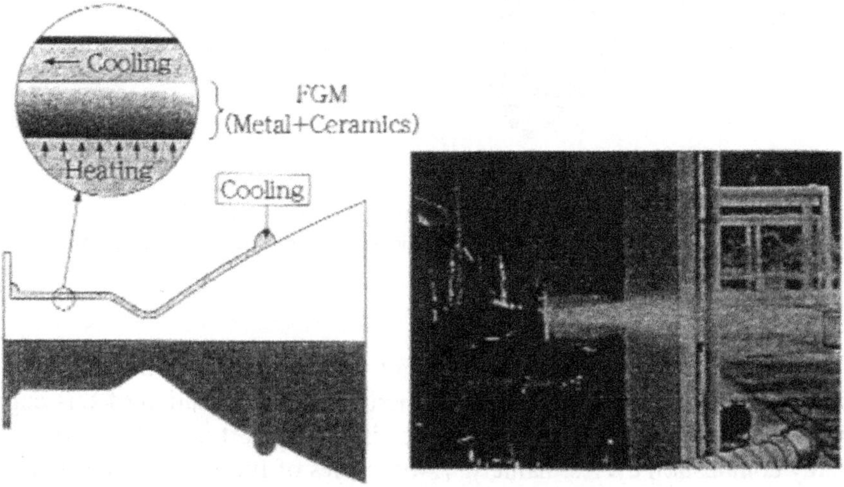

Figure 5.30. The construction of the rocket engine and combustion test.

3.3.2 RCS Engine for HOPE

Engines of space planes must have high heat resistance and minimum weight. Carbon Composite Materials (or C/C materials) are thought to satisfy these criteria. To use C/C materials in an oxidizing environment, they must have oxidation-resistant protective coatings such as SiC (silicon carbide). However, due to the difference in the thermal expansion coefficients of the SiC and C/C materials, the conventional coating method often results in delamination of the coating material from the C/C layer in high temperature, such as in the rocket combustion chamber, thus causing problems with reliability and durability. Niino *et al.* used SiC/C FGM for the RCS engines of HOPE with a view to improve the durability and relax thermal stress. For the substrate C/C material to be airtight, the procedure of pitch infiltration, hot isostatic pressing and high temperature sintering was repeated eight times to densify it. The 0.04-mm thick graded layer of SiC/C was formed by chemical vapor infiltration (CVI). Within the composition-controlled source gas mixture of tetrachlosilane-methane-hydrogen ($SiCl_4$-CH_4-H_2), the $SiCl_4/CH_4$-gas ratio is changed with time by the computer, thus changing the composition rate of the layer from C to SiC continuously.

Finally, the SiC layer was formed by chemical vapor deposition to make the whole layer 0.1 mm thick. Figure 5.31 shows the structure of the engine.

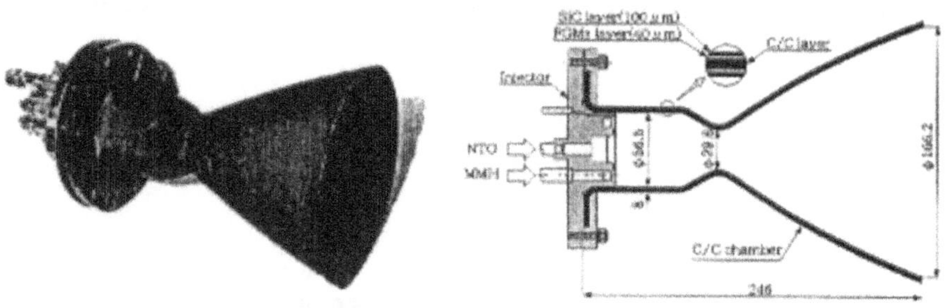

(a) (b)

Figure 5.31. Structure of the engine.
(a) The RCS thruster used In this test, (b) the designed thruster using a steel injector body and a film cooled C/C-SiC FGM chamber.

The RCS engines are used in orbit to control the vehicle's position, burning the fuel for short intervals. Therefore, the evaluation of this engine was carried out by pulse combustion testing with short combustion time and many combustion cycles, namely, 7,465 cycles of fuel combustion in a total of 7465 seconds. When FGM was not used, cracks between the C/C layer & SiC layers occurred after a single test cycle, but when FGM was used, it withstood 900 cycles before delamination was observed. This shows good prospects for the engine's practical applications.

3.3.3 FGM Returns from Space

HIPMEX (High Performance Material Experiment) is a part of NEDO's (New Energy and Industrial Technology Development Organization) Research Project on Super Refractory Materials in which SiC-C/C FGM is actually used as the thermo-protective material of the capsule and is exposed to the atmosphere when the capsule reenters the atmosphere on its return from space[14]. On January 15th, 1995, M-SII launched this type of capsule at the Kagoshima Space Observation Center but failed to orbit it, and the capsule was retrieved from space in July 1996. Detailed data on this experiment are not available, so one must speak in terms of estimation: Supposedly the greatest thermal loading was 1450°C for approximately 10 min. As a comparison, four other materials made of SiC and C/C materials (which were not FGMs) were launched with the capsule, and after being returned, their condition was compared with that of the SiC-C/C FGM. The soundness of the material was best preserved in the FGM of SiC-C/C, and microscopic examination shows the SiC layer to be almost sound and reusable. There is damage at the base, but other specimens also bear the same visible damage so external forces on Earth after it was collected probably caused them.

3.4 Ultra-Light TPS and FGM

The demand for high insulating efficiency is greater for TSTOs because of the longer length of time spent in the atmosphere, therefore a particularly large amount of heat-insulating material has to be used as shown in Figure 5.32 by using conventional TPS design.

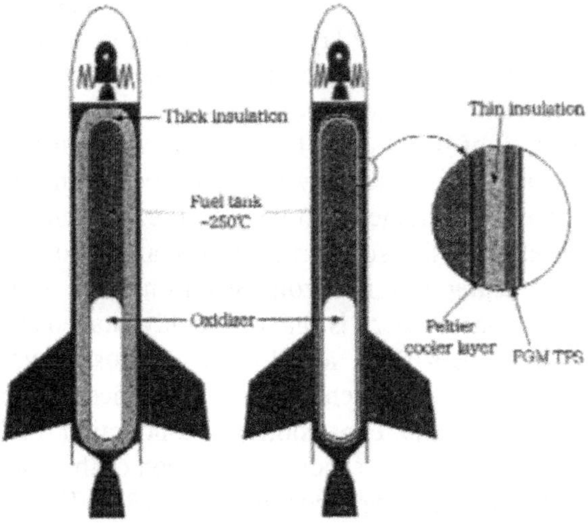

Figure 5.32. Super lightweight body construction.

By applying FGM technology, TPS is expected to have ultra-high insulating efficiency to resist temperatures up to 2,000°C. This technological innovation will greatly reduce the thickness of TPS and thus significantly lighten the body weight of the vehicle. Among the candidate TPS materials to be used in place of the conventional ceramic tile is thermo-resistant FGM coatings on the surface of the vehicle body, the invention of Rocket Dyne Ind. in the US.

Furthermore, Kawasaki *et al.* have worked on improvement of oxidation resistance by applying the FGM concept, and they have PSZ/Al_2O_3/NiCrAlY as an example of a functionally graded oxidation-resistant TBC. Their results show that durability was considerably improved with the FGM coatings compared to the conventional two-layer coatings.

3.5 Long-Life Engines and FGM

RLVs require long-life engines because they are not disposed of, as are conventional engines. Here again, radically new technologies and ideas are needed. The essential elements in designing these long-life engines are the design of thermal barrier coatings (TBC) that block the heat flow from the 3,000°C combustion flame into the combustion chamber wall and the design of the compliant structure that relieves the thermal strain of the combustion chamber[16].

The temperature of the wall surface of the inner cylinder made of copper alloy, which faces the combustion flame, must be cooled down sufficiently to withstand a few hundred firings of the engine. To achieve this under a heat flux that reaches up to 100 MW/m^2, it is considered that the cylinder wall needs a 30 μm thick ZrO_2 coating layer. In addition, it is considered that the coarse interface is the primary cause of delamination of the coating layer from the substrate, and that the layer should be processed on a smooth surface. Also required of the coating layer is high thermal shock resistance, because vertical cracks are induced by thermal shock at the shutdown of the engine. For the above requirements to be met, i. e. the thinness of the layer and its tight adhesion to the substrate, the research program began on a processing method using the Electron Beam-Physical Vapor Deposition (EBPVD) technique. Shown here is one of the advantageous features of the coatings produced by EB-PVD, a columnar microstructure. Since each column-shaped crystal is independent of the other crystals, they can separately follow either the expansion or contraction of the substrate material, supposedly resulting in efficient relaxation of thermal shock.

Figure 5.33 shows a prototype coating with EB-PVD method. The cup test to measure the mechanical adhesion of the coating to the substrate is being carried out with the cooperation of the Hirai Lab of Tohoku University. Also scheduled is an evaluation of thermal shock resistivity, using the high power YAG pulse laser to simulate the high heat flux of 100 MW/m^2 in rocket engines. The results of these evaluation experiments will be the basis upon which we will continue to improve the resistance of materials to the combustion gases of the high-pressure staged combustion rocket engines.

At the National Aerospace Laboratory cooperating with MHI Niino *et al.* also started investigation of a structure with the outer shell made of C/C composites that is expected to have the effects of a low stiffness structure because C/C composites are light and their thermal expansion coefficient is very low. As a result of model analysis, by controlling the stiffness of the outer cylinder made of C/C composites, the longevity of the inner cylinder made of Cu alloy was doubled compared to the metal shell made of Inconel.

The prolongation effects on engine life expectancy by TBC and the compliant structure can be multiplied.

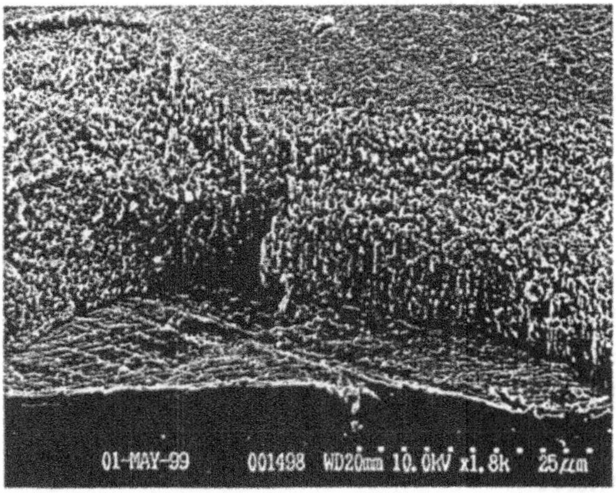

Figure 5.33. Prototype coating with the EB-PVD method.

3.6 Health Monitoring and FGM

Both the engine and the body of a reusable vehicle must have health monitors implanted in critical parts to have safety guaranteed after hundreds and thousands of recycles.

To maintain the soundness of the structures, RLVs must undergo quantitative damage detection programs, and need to be repaired or have parts replaced according to the extent of the damage incurred. This is called health monitoring and is another important aspect in our goal to establish reusable vehicles. Sensors to monitor the condition of the vehicle parts are under development. Among them is a thermal sensor made from a thinned thermoelectric converter that should be able to detect temperature rises due to damage in parts such as the thermal barrier layer and the bearings of the turbo pump. FGM is thought to be effective in relaxation of thermal stress around the spots where the sensors are inserted, so that the sensors themselves are protected from thermal expansion of the surrounding material.

In the United States, the FRP-produced extremely low temperature tank of X33 is monitored by a network of fiber optic materials all over the structure [17], to detect any local fracture or stress concentration. Again the FGM concept could be applied to fiber optic materials to produce higher efficiency.

3.7 Other Possible Applications of the FGM Concept

The FGM concept is favorable for the component control of solid lubricant materials to bind the body structure and surface lubricant as tightly as possible.

The bodies and engines have many extremely hot spots with coolant adjacent to them. At such spots where heating and forced cooling occur at the same time, large temperature differences can be used as a power source by inserting FGM thermoelectric converters.

Another example is debris protection materials, which are usually made from one component. But if the material can have functional varieties, namely, both shock absorbing properties and toughness as an integral part of the structure, damage caused by debris in space could be protected to a large extent.

3.8 Summary

Niino *et al.* have tried to offer a few suggestions that they think may contribute to the design and construction of new types of hypersonic space planes, RLVs. To develop RLVs, many more approaches for adopting the FGM concept must be discussed. Second-generation space planes must be developed within a relatively short period of time to be competitive in the world market, while we strive to reduce costs and establish reliability. Japan is fortunate to have accumulated considerable findings and expertise from numerous projects; namely, the project that started in 1987 under the Science and Technology Agency, the other FGM project from 1995 under the Ministry of Education, NEDO's Research Project on Super Refractory Material, and finally, the large FGM fabrication project under the leadership of Mechanical Engineering Laboratory of Industrial Science and Technology. These research projects are not limited to the area of thermal stress relaxation, but extend to functional materials such as an energy converter, the application of which has a good potential for various uses such as the health monitoring sensors. A wide range of technology must be applied for the RLV to be realized so cooperation beyond borders from governmental, academic or industrial organizations is hoped for. This fiscal year our Functionally Graded Materials Forum will start investigating further research areas and potential applications for this field in response to NASDA's request. Any form of cooperation will be welcomed.

REFERENCES

1. Tsuda, K., Ikegaya, A., *et al.*, Powder Metallurgy, 39 (1996), 296-300.
2. Ikegaya, A., *et al.*, *Functionally Graded Materials*, Elsevier Publishing, (1996), 361.
3. Ikegaya, A., *et al.*, *Functionally Graded Materials*, Trans Tech Publications, (1998), 487.
4. Ikegaya, A., *et al.*, EURO PM '99 Conference on Advance in Hard Materials Production.
5. Suehiro, Y., *et al.*, Society of Petroleum Engineers International 2000, 59423 to be published.
6. Tokita, M., J. Soc. Powder Tech. Japan, 30-11 (1993), 790-804.
7. Enomoto, Y., Ichikawa, K., Tokita, M., Dapkunas, S. J. and Smith, D. T., Proc. Cooperating Internationally: The US.-Japan Civil Industrial Tech. Arrangement Ann Arbor, Mich. (1998), 8-1.
8. Tokita, M., Development of Large-Size Ceramic/Metal Bulk FGM fabricated by Spark Plasma Sintering, Proceedings of the 5th International Symposium on Functionally Graded Materials FGM 98, Dresden, Germany, (1999), 83-88.
9. Niino, M. and Koizumi, M., MRS Bulletin, Overview of FGM Research in Japan, 20-1 (1995), 14.
10. Research and Development Bureau of Science and Technology Agency, Report from "Conference on Future Space Transportation System" (in Japanese), (1999).
11. Foundation for Promotion of Japanese Aerospace Technology, Report from "Forum of Visionary Laser Energy Network" (in Japanese), (1999).
12. Kuroda, Y., *et al.*, Proceedings of the 1st International Symposium on Functionally Graded Materials (1996), 469.
13. Kiuchi, N., *et al.*, Proceedings of the 5th Japan International SAMPE Symposium, (1997), 1165.
14. Kiuchi, N., *et al.*, Proceedings of Functionally Graded Materials, (1997), 95.
15. Sakai, T., *et al.*, Proceedings of the Autumn Session of the Japan Institute Metals (in Japanese), (1999).
16. Moriya, S., *et al.*, Discussion on Life Prolongation of Rocket Combustion Chamber, Processing of the 40th Air breathing engine and Space propulsion Conference, Jap,. 2000, (in Japanese), (2000), 187.
17. Bodan, P. and Bouvier, C., AIAA/ASME joint Conference for Structures, Structural Dynamics and Materials, (1998).

Index

GPSR Compliance

The European Union's (EU) General Product Safety Regulation (GPSR) is a set of rules that requires consumer products to be safe and our obligations to ensure this.

If you have any concerns about our products, you can contact us on ProductSafety@springernature.com

In case Publisher is established outside the EU, the EU authorized representative is:

Springer Nature Customer Service Center GmbH
Europaplatz 3
69115 Heidelberg, Germany

Batch number: 10755292

Printed by Printforce, the Netherlands